PREGNANT
BEARS
&
CRAWDAD
EYES

PREGNANT BEARS & CRAWDAD EYES

Excursions & Encounters in Animal Worlds

Paul Schullery

The Mountaineers/Seattle

5 4 3 2 1
5 4 3 2 1

Published by The Mountaineers
1011 S.W. Klickitat Way, Suite 107, Seattle, Washington, 98134

Published simultaneously in Canada by Douglas & McIntyre, Ltd.,
1615 Venables Street, Vancouver, B.C. V5L 2H1

Published simultaneously in Great Britain by Cordee,
3a DeMontfort Street, Leicester, England, LE1 7HD

Manufactured in the United States of America

Edited by Barry Foy
Cover design by Elizabeth Watson
Cover art and interior illustrations by Jim Hays

Library of Congress Cataloging-in-Publication Data

Schullery, Paul.
   Pregnant bears & crawdad eyes : excursions & encounters in
animal worlds / Paul Schullery.
      p.   cm.
   ISBN 0-89886-293-0 (cloth). — ISBN 0-89886-292 (paper)
   1. Animal behavior. 2. Animal ecology. 3. Human-animal
relationships. 4. Evolution (Biology) 5. Animals—Anecdotes.
I. Title. II. Title: Pregnant bears and crawdad eyes.
QL751.S348   1990
591.5 ' 1—dc20   91-11164
                                                        CIP

For my mother

For the animal shall not be measured by man. In a world older and more complete than ours they move finished and complete, gifted with extensions of the senses we have lost or never attained, living by voices we shall never hear. They are not brethren, they are not underlings; they are other nations, caught with ourselves in the net of life and time, fellow prisoners of the splendour and travail of the earth.

——Henry Beston, 1928

Table of Contents

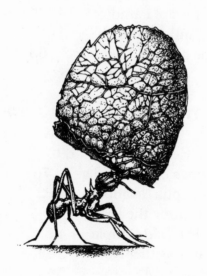

Introduction:

Savoring the Questions

About twenty years ago, a friend and I (or, as the song says, "me and this *other* fool") drove from Ohio to Panama and back. It's a long story—it's a long drive—but one brief part of it comes to mind as I reread these essays.

We were in Guatemala, having entered it from Belize across a frontier that might best be described as uneasy. The Guatemalans even today are resentful that Belize exists, and if little Belize lost the military protection of big England, the Guatemalans would probably absorb it very quickly.

Once into Guatemala, things didn't get all that much better. The local politics were uncertain enough to keep us nervous; being pulled over at gunpoint by teenage soldiers may be old hat for globe-trotting journalists, but it was serious stuff to us small-town Ohio boys. Anyway, our senses were fairly well tuned up, which may explain why I noticed the ants.

I was driving our dusty little white Beetle along the dirt road (I can't say we were off the beaten path, because that's exactly what the road looked like—a beaten path) when I suddenly slammed on the brakes and skidded to a stop. I remember that as I started to back up, my pal got a little annoyed—we were making bad enough time as it was—but as soon as he saw what I had seen, he was as excited as I.

There was a double line of large ants crossing the road, most of them carrying pieces of vegetation several times their size. At the time I knew only that they were ants; I later learned that they were probably leaf-cutting ants, known to the local Maya as "wee-wee" ants. Not knowing their name didn't make them any less exciting.

We watched them for a few minutes, bending down close to see how amazingly large their loads were. We wondered what they were going to and coming from in the jungle, decided not to look, took some pictures, and drove the car over them once more as we left. Ohio, I thought at the time, had nothing like this.

I was wrong, of course; Ohio had very similar things. The difference—what Ohio didn't have—was the setting. Seeing a little line of ants crossing a dusty spot in the lawn under the maples was nowhere near the thrill of seeing that same procession passing between two wildly tangled walls of jungle. It was the stage, not the players, that made the drama so exciting. Anything one sees far away, in an exotic place, while concerned about automatic weapons, tends to take on a good bit of romanticism in the memory.

I'm as charmed as anyone by the rare and exotic. I read a lot about African and Asian wildlife, and the grizzly bear has long been a special interest of mine. But I must admit that most of the questions that have engaged my attention in natural history have been inspired by animals we tend to take for granted.

It is a truism among naturalists that there is just as much wonder to be found nearby as far away, if only we look for it and see beyond the surficial familiarity. In this we all can take our lead from Louis Agassiz, who once wrote, "I spent the summer traveling: I got halfway across my back yard."

As you read this book you will notice that I've had several backyards, east and west, north and south, and that I've traveled in other people's backyards as well. I don't want to appear more a homebody than I am; depending upon where you live, some of the animals I talk about here might seem exotic to you. When I lived in Vermont and Pennsylvania, none of my friends were acquainted with elk, or horned lizards. These friends might have smiled in sympathy at a western friend of mine who was in her late twenties before she came east and saw her first cardinal.

On the other hand, when I first visited the Pacific Northwest, I was surprised to discover that most people living along salmon streams thought no more about those spectacular fish and their amazing migrations than most Midwesterners do about the carp in the local pond. We seem able to take almost anything for granted, but are most likely to underappreciate that which is nearby. The backyard is a neglected recreational resource.

For convenience' sake, I have arranged these essays in three groups. I recognize some artificiality in this arrangement and invite you to read them in any sequence you like, but there is some comfort in structure, too.

The first group, called "Staying Alive," seems to me to focus most directly on the process of survival, either as individuals or as species. You will see that my definition of survival is pretty broad, covering everything from birth storms to the animal subconscious. But I at least see a thread here, of how animals deal with or perceive their surroundings, and how they make the best of what they have to get by.

The second group of essays, "Wild Society," is pretty much about how animals deal with each other, which comes down to how they share their habitat. I take my cue from the ecological disciplines, which over the past few decades have switched their heaviest emphasis from the study of populations of individual species to the study of communities of interacting species. The more time I spend in nature and

11

learning about nature, the less I spend thinking about individual animals. In short, what I'm most concerned with here is cooperation and competition, which, I think, sometimes turn out to be parts of the same process.

The third group of essays, "By Any Other Name," concerns itself with some of the perceptual filters through which we see animals. Every time I remove a filter, I find myself looking through another one. Some of these filters trouble me as inappropriate, some seem mightily helpful, and some interest me just because of what they tell me about myself and my fellow humans.

As I reread these essays, it occurs to me that I don't really expect ever to fully understand my relationship with wildlife. If all I sought were biological data, I would have hopes for a fairly complete understanding. But, like so many observers before me, I'm looking for more than that. Coming to terms with other animals is much more than learning how they survive, or why they behave the way they do. Though I have great enthusiasm for all the engaging things that science can tell me about animals, at times I prefer less quantitative views that let me see them as something other than study specimens, and that give them the kind of respect Henry Beston had in mind when he called them "other nations."

Regardless of which view I take at a given moment, I also recognize that one of the great pleasures of my relationship with other animals is in the process of learning about them, not for the sake of the information gained but for the joy of the search. As much as anything else, this book is a celebration of the gentle adventures I have had while searching. Usually I do want an answer, but whether I find one or not, I always enjoy savoring the questions.

Paul Schullery
Yellowstone Park
February 1991

PART I

STAYING ALIVE

JIM HAYS

Starker's Doves

Starker Leopold trapped doves on his back porch. He lived in a handsomely landscaped home in Berkeley, near his office at the University of California, and he set up a simple trap much like the ones children use to catch robins, rabbits, and squirrels. A stick held up one end of a box, and a string attached to the stick led into the house, where Starker waited to pull it when a bird went underneath to get the bait.

It was a delight and a lesson to me to discover that he did this. The first time I visited his home—about 1979, I guess—he took me out and showed me the trap. By then he'd trapped, leg-banded, and released more than a thousand doves and was working on his analysis of the returned leg bands.

The memory of seeing Starker working his little trap has since been the source of much happy inspiration to me. Here he was, an august personage in conservation and ecology, author of several superb books—including bona fide classics on the wildlife of Mexico and the California quail—and scores of important articles. One of the leading figures in post–World War II wildlife issues, chief author of the famous "Leopold

17

Reports" in the 1960s, which set the course for so much of federal wildlife management, Starker did profound honor to the legacy of his father, Aldo Leopold, now regarded as the father of modern wildlife management in North America. He advised nations on wildlife. He traveled everywhere to study ecological systems and issues. In short, he dealt with nature on a global scale, yet he never lost the curiosity that made him such an enthusiastic and successful scientist in the first place. There were a lot of doves at the feeder out back, and no one seemed to know much about their life history or movements, so why not crowd another little study into the busy days? Won't take that much time, will it?

He learned many things from the study, but the one that struck me when he mentioned it that day was that the two biggest killers of doves in his area were cars and cats. His doves weren't as heavily hunted as many dove populations, spending as much time as they did in suburban and urban areas, but they had about the same life expectancy as heavily hunted doves. They simply had different kinds of hunters to contend with. Starker seemed as pleased to have discovered all this as if it had been his first undergraduate project. That love of information, whether it was of international value or merely illuminated some aspect of his neighborhood, is what I remember best about him.

A few years later, my wife and I were sitting at our kitchen table talking, when she glanced out the window to the backyard and gasped, "My God, there's a cat!"

My first look took in the whole sad story. We'd set up a sunflower-seed feeder on a low pole, a good five feet from the nearest hedge. The feeder was popular with the local grosbeaks, who consumed unseemly quantities of the expensive seeds, earning the nickname "the piggies" from my wife. As summer passed, however, the scraggly brush along the fence flourished, spreading generously in all directions, including toward the feeder. Now a neighbor's cat had climbed the thick

18

brush to the height of the feeder, and was launching himself at the crowd of preoccupied grosbeaks.

I was out the back door almost as fast as the cat's jump, in time to see him, a male grosbeak gripped in his teeth, vault the fence and race away. As I sprinted down the alley after him, I could see bright yellow and black feathers jutting from either side of his face.

The cat easily outran me and dashed into another yard; I didn't even get the satisfaction of throwing a rock. I was furious, but even then I knew it was hard to blame the cat. His idiotic owners were the real culprits, and the town was largely populated with such people—stupid, well-meaning "animal lovers" who let their dogs and cats run free day and night to dig, defecate, howl, and kill at will.

Actually, you kind of have to admire the modern cat. After thousands of years of human efforts to turn it into a piece of ornamental furniture, it can still go out and make a living. I don't think we ever entirely domesticate *anything,* but we haven't even made a good start with cats.

I wasn't too anxious to admit that some of the blame was mine, too. I knew the hedge was too big and needed trimming, so I suffered the twinges of conscience many bird-watchers feel when they realize that by inviting birds to gather in their yards, they are also feeding bird predators. After all, Starker had warned me.

Shortly after Starker's death in August of 1983, I went to the library to look up his dove paper in the *Journal of Wildlife Management.* There, with appropriate graphs, tables, and maps, were the results of his experiment in backyard biology. There was even a nice picture of the trap, which was described as follows:

> The trap consists of a hardware-cloth box, roughly 56 × 41 cm., hinged to the tray. Commercial bird-seed served as bait. A stick propping up the trap

can be pulled out with a string extending into the house, dropping the trap....

The usual rule when describing research techniques in these journals is to give a parenthetical citation of the author and date of the publication that first described the technique, as in "Data were analyzed using the Frobisher Model (Frobisher and Ferguson, 1952)." I wondered if Starker had been tempted to conclude his description of the trap with "(Little kids everywhere, all the time)."

For me, Starker's doves have gradually come to symbolize the complex interweavings of the two idealized worlds so often presented in nature writing: civilization and wilderness. Many writers have casually commented on the differences between the two worlds, giving each some popular stereotype, as in "Civilization is dirty and wilderness is pure."

The smartest observers (and Starker was surely one of those) long ago realized that there are not two worlds, but one world with an amazingly involved continuum of conditions, from the almost completely sterile and artificial constructions of human technology at its most powerful to the almost totally unmanipulated wilderness setting.

Of course, those same observers also realized that, set upon this spectrum at a sort of oblique angle, is another, a spectrum of opinions (most quite old) about how valid it is to use such terms as "artificial" to describe the works of humans. At one end of this spectrum are those who see human activities as qualitatively indistinguishable from the activities of other life forms: Birds build nests, ants build hills, humans build Astrodomes. At the other end are those who consider all human acts to be separate from natural processes: We are guests in the natural world. Between the two extremes lie all manner of other opinions, grading into one another, about just where humans fit into nature. Add to this, too, a complementary spectrum of opinions that address such questions as whether or not what humans do to nature, or as part of nature, is good, bad, or neutral.

These opinions are central to modern debates over care of the environment. Two extreme positions I've seen in recent years provide some example of how far apart we are as a species in our view of our world. Famous scientist and nature philosopher René Dubos writes in a small book, *The Resilience of Ecosystems* (1978):

> Nature is like a great river of materials and forces that can be directed in this or that channel by human interventions. Such interventions are often needed because the natural channels are not necessarily the most desirable, either for the human species or for the Earth. Nature often creates ecosystems which are inefficient, wasteful, and destructive. By using reason and knowledge, human beings can manipulate the raw stuff of nature and shape it into ecosystems that have qualities not found in the wilderness. They can give a fuller expression to many potentialities of the Earth by entering with it in a relationship of symbiotic mutualism.

I have friends who would say this is so much anthropocentric hogwash; that Dubos is presuming vastly on nature by claiming we know what is best for it. I personally would find it much easier to argue against Dubos' reasoning than for it. But there are a lot of people who happen to go along with this view.

So try an opposing view, which other friends of mine would find just as offensive. Biologist David Graber, in a review in the *Los Angeles Times* of the book *The End of Nature* (1989), by Bill McKibben, writes:

> Human happiness, and certainly human fecundity,

21

are not as important as a wild and healthy planet. I know social scientists who remind me that people are part of nature, but it isn't true. Somewhere along the line—at about a billion years ago, maybe half that—we quit the contract and became a cancer. We have become a plague upon ourselves and upon the Earth. It is cosmically unlikely that the developed world will choose to end its orgy of fossil-fuel consumption, and the Third World its suicidal consumption of landscape. Until such time as *Homo sapiens* should decide to rejoin nature, some of us can only hope for the right virus to come along.

The result of this multilayered storm of rhetoric, scientific inquiry, and religious conviction is that the very words "natural" and "unnatural" have long been hard to define or use to the comfort of everybody, and the real world, in which all the elements of the setting are interacting in such involved ways, becomes quite a challenge for the careful thinker.

Another day in my kitchen, I looked out into the backyard just in time to see all the birds scatter as if a grenade had gone off in the feeder. A small hawk was just completing its dive and swooped low over the lawn before flying away empty-taloned. This gave me a different set of twinges. I was, in fact, excited to see the hawk—to know it was in the neighborhood, contributing to the diversity and dynamics of local wildlife. I wouldn't even have minded if it had got a bird. But again, I knew I'd set the birds up for it by attracting them into the open. Somehow, because the hawk was a wild-born predator, a native of the area, and a part of the pre-European-settlement life cycles of these birds, I didn't mind its hunting as I had minded the cat's. But I knew all along that neither the hawk nor the cat was engaged in an activity whose causes or consequences were simple.

Starker's doves live in a busy world. Eating human-provided food, any one of them might die at the claws of a creature that is the result of an extraordinary combination of human manipulation and the fierce persistence of millions of years of behavioral evolution. The dove builds its nest of vegetation, some wild, some human cultivated, some growing against the will of humans but in sites made hospitable to it by human disturbances. The dove eagerly exploits every opportunity its crossbred environment provides. It is not tame, but it is deeply habituated to humans. It is wild, but it does not live in a wilderness. It lives in a world of compromises, where humans and other life forms carry on a series of holding actions. A man edges his sidewalk, an ant colonizes the man's kitchen; the doves escape from hunters, but fall victim to cars and cats.

I think of this richness of interaction, of the elegant textures and limitless potential directions it affords, and my little backyard seems an exciting place, where there is no end of adventure and fascination, and where I can admire and consider all the kin of Starker's doves.

Embryonic Journeys

In my natural history readings I wander among several literatures, often wading determinedly through the most technical scientific monographs, but more often slipping lazily into that vast and troubled body of popular publication known by scientists as "the gray literature," or, less kindly, "the throwaway press." Often, the best the throwaway press has had to offer has been in its older forms, especially the older nature and sporting books and periodicals—say, before 1920, back when an "outdoor magazine" might contain as much nature study as it did hunting and fishing.

I must admit that sometimes it seems that I go about my informal education all backwards; it might take great amounts of reading this stuff before I stumble upon a really good question. But it has taught me that finding the right questions is just as important as then learning their answers. After all, most of the questions an amateur like me might have about nature were almost certainly asked long ago by thoughtful scientists. If all I wanted were answers, I'd just read the science in the first place. But there's nothing quite like stumbling around in the dark with previous generations of amateurs to teach you how we got to know what we know today.

And so it is that I rarely hesitate to invest time in old hunting, fishing, and adventure books, and in great old periodicals like *American Naturalist, Forest and Stream,* and even the venerable *American Turf Register,* whose distinguished career ended long before the Civil War. For my purposes, the older the better. I don't especially care what we misunderstood recently; I want to know how we got it wrong in the first place. I want to trace our education from abject ignorance all the way up to modern times and enlightened, reasonably well-informed ignorance.

Take bears. Within a few miles of where I'm sitting as I write this, any number of large male bears—both black and grizzly—are in the last few weeks of their winter denning. Not long after they emerge, single subadults and then females with young will also appear. I'm only one of many writers to celebrate the extraordinary evolutionary achievement of winter denning; still, we haven't said much about one of the most intriguing sidelights of winter denning, perhaps because science hasn't given us complete answers yet.

I'm talking about delayed implantation, a little-known and even less-understood phenomenon that suggests just how flexible mammalian physiology can be. Delayed implantation is a process by which a newly fertilized egg halts development at what is called the blastocyst stage, when it's a little bundle of cells, and floats freely in the reproductive tract of the animal for an extended period of time. After this time—whatever time works best for the animal in question—the egg implants itself in the uterine wall, resumes normal development, and the fetus then continues its growth until birth.

I came upon indirect mention of delayed implantation when I was researching the first of my three books about bears. In the early 1800s an American naturalist observed that "no man, either Christian or Indian, ever killed a She-bear with young." It's this sort of little aside, a mere footnote in some forgotten narrative, that speaks volumes of mystery to the alert reader: What's that—nobody ever killed a pregnant female bear? How could that be? What are the odds that all those thousands of hunters would only kill barren or unimpregnated sows, never once finding a partially developed fetus? It's not possible.

But of course it is; that's just what happened. And why it happened is a satisfying, if occasionally puzzling, example of how richly layered with questions the ecology of an animal can be.

Black and grizzly bears usually mate in late spring and early summer. But implantation is immediately delayed, and the blastocyst remains more or less invisible to human vision until November or so, when implantation occurs, just about the time the bear is preparing to den. Then the sow experiences a fairly short active gestation period for such a large animal, and the cubs are born in midwinter, say, late January. They weigh a pound or less at birth and grow to five or ten pounds by the time their mother leads them out of the den in April.

So, lacking microscopes to examine the reproductive tracts of dead bears, and having no contact with the few scientists who even then had a notion that delayed implantation might be occurring in some animals, our naturalist forefathers assumed that the sows they killed and butchered were not pregnant. The egg was much too small to see.

It's been fun tracing this question through progressive ages of nature writing. Some writers were practical about it: having no evidence that bears were pregnant before winter, they announced that bears must mate in the fall. Others, knowing darn well from their own observations that bears mated in June, simply fell silent on the whole matter. I think others avoided it out of Good Taste, not really wanting to deal with the sexual shenanigans of the beasts in the first place.

But let's not get smug yet. Delayed implantation is now

well documented, but biologists still have a lot of questions. When I started reading about delayed implantation, I had a few myself. I could see why it made sense for the sow to give birth to very small young. After all, she was stuck in that den, and if she gave birth to cubs that had developed through a full seven- or eight-month pregnancy, they would drink her dry of milk in a matter of days. But with the cubs born weighing only a few ounces, their milk needs were manageable, and they could continue their fetal development, so to speak, outside the womb, being essentially fully grown "newborns" when the family left the den.

But there are other questions of timing. Why not just mate in the fall and let the egg implant immediately? Well, I'm not sure, but I imagine it has to do with the need to get the very time-consuming and nutritionally draining (for the male, especially) business of mating out of the way early in the year, before the bear's metabolism goes into high gear and it begins to put on heavy fat for winter. Late summer and early fall would be a terrible period to interrupt with a mating season; that's when bears are working on those narrow margins of overeating that will get them through the winter, and it's when some of their richest food sources are available.

That answer doesn't entirely satisfy me, though. The elk, deer, and bighorn sheep in my neighborhood all mate in the late fall, and though it's just about all the males can do to struggle through the winter after all that exertion, some make it, and the arrangement works just fine for those animals. I do suspect, however, that good bear foods—which include those very same elk, deer, and sheep—are available later into the fall than good grasses and browse, so fall probably offers the bear more nutritionally than it offers the vegetarians.

As well, those browsers and grazers don't den up. They keep eating all winter, and though they don't thrive on the dried plant matter they dig out of the snow, it does help slow down their weight loss and physical deterioration. The bear

27

completely stops eating, and so every late-summer mouthful counts.

But on to other questions. Why not just mate back in June (if you're a bear, I mean), delay implantation until fall, and *then* carry the fetus through a full-term pregnancy and give birth in spring? That way, you'd avoid the whole problem of feeding cubs in the den. Well, spring is a pretty lousy time for bear food in a lot of places, and many bears take the first few weeks to get into the habit of eating heavily again. That would be an inopportune time to go through the stress of giving birth to large and very hungry cubs, I suppose. No doubt much important bonding occurs during those two months the cubs spend in the den with their mother.

Maybe there are physiological limits to how long implantation can be delayed. Maybe it's no coincidence that implantation seems to occur just about the time denning occurs; maybe the remarkably involved metabolic processes associated with hibernation put certain limitations on what the reproductive system can get away with. Maybe some scientist whose work I've missed has figured this one out and I'm stumbling around in unnecessary ignorance just like my nature-writer forebears. I'll watch my mail.

The answers will come, I'm sure, from many directions. Delayed implantation occurs in many mammals; a partial list includes several bats, badgers on three continents, fisher, mink (because they're raised in captivity for furs, they're probably the most closely studied of all), skunks, more than a dozen species of seal, batches of mice and rats, shrews, moles, roe deer, and the armadillo, which may hold the record, being able to delay implantation for well over a year. What's more, they all do it in different ways.

Some, like the bears, have a seasonal cycle that's pretty well fixed: spring mating, summer delayed implantation, fall implantation, for example. Others, like some of the mice, are known to breed actively over an extended period of time, so delayed implantation seems to kick in if environmental stress "tells" the animal's system that it would be a bad time to develop a fetus. The female red kangaroo may carry an unimplanted blastocyst at the same time that she's nursing a joey in her pouch; if something happens to the joey, the blastocyst seems to serve as a backup unit.

Small animals typically have very short gestation periods; it appears that delayed implantation may be the only way they can arrange to give birth in the optimum season, especially if that season occurs during or at the end of an extended period of environmental stress, such as winter. One species of bat appears to have developed a slightly different form of delay, in that the growth of the egg does not completely stop; it just goes very slowly, gradually accelerating late in the fetus's term.

I've given you a sort of *Reader's Digest* Condensed Essays summary here. The scientists I've been reading are deep into discussions of the chemistry of all this, and the subtle effects of changing photoperiods, and lots of other rewarding things. The questions lead to more questions. There are bats who have taken a whole different approach: they mate in fall and winter, but they delay fertilization rather than implantation. The females store live sperm in their uteri for up to ten weeks before ovulating, so that the sperm have something to fertilize. Imagine the convenience. Romantic hedonists have it that variety is the spice of life; whether or not that is true, variety is surely the essence of reproductive biology.

Jeremy Schmidt, a travel-writer friend of mine, recently wrote that "the earth works wonders and holds them secret." He might have added that it's all the same to the earth whether or not we ever figure them out, or even if we notice. There are always new secrets, but the harder we look, and the more questions we answer, the more obscure and hazy the new secrets become.

Perhaps that's another reason I spend so much time

seeking out the old questions. The biologists now studying delayed implantation are asking questions so technical that I can barely follow them.

Perhaps that's also why we hear so little about delayed implantation in today's magazines and books; questions of physiology and biochemistry don't have the public appeal that more visible animal traits do. I'd hate to think, though, that one of the nineteenth century's most intriguing wildlife questions—Where are all the pregnant bears?—should reach a dead end of public interest just because the answer isn't simple enough, or easily enough illustrated with color photographs in a nature magazine, to appeal to us today.

I'd better let it go at that, before this essay degenerates into a polemic on the growing crisis of scientific illiteracy. I think that what the discovery of delayed implantation teaches us, and what Jeremy was really saying, is that the earth works wonders and holds them secret only until we have enough pride and gumption to try to figure them out.

Chickadee Down

Wherever I have wintered in the north country I have been accompanied by chickadees. There are usually other birds about, but for some reason, probably having to do with their exuberant manner, the chickadees always come to my attention the most. Though it be twenty below, there are still small flocks of them batting here and there in yard and woodlot, chattering like second-graders at the first spring recess.

They never seem to stop; even when perched they engage in little stationary flits, changing their grip, flipping their tails, always active. They are the only winter bird that makes me wonder about avian metabolism, because theirs must be amazing to let them do what they do. If I were to take a small chunk of glowing coal about the size of an adult chickadee, and put it on a chickadee perch, it would be dead cold in ten minutes. And yet this little bird, through judicious and enthusiastic stoking of its much more modest furnace, and also through some combination of other physiological mysteries, goes on winter after winter, for as many as ten or twelve years.

Much of what birds do to survive the cold is obvious. They hole up, for one thing. Chickadees are among the spe-

31

cies that sit out the nights—darkness being the hardest time because it is coldest and least practical for food gathering—in shelters of various kinds. The best shelter is usually a tree cavity or other enclosure, but a study of redpolls in Alaska showed that the redpolls (one of the chickadee's few real rivals for arctic survival among small birds) sheltered effectively from the cold in white spruce, whose dense needles and boughs are almost the equivalent of a hollow tree trunk in protective qualities.

There are some peculiar risks associated with such shelter, by the way. Swallows studied in Canada would "cluster" in a good tree cavity, jostling for the warmest spot inside the hole; but if birds on the outside, closest to the weather, succumbed to the cold, the ones that had won the warmest spots were doomed, trapped under the press of dead companions.

Most birds are also more warmly dressed in winter. We usually think only of mammals when we hear of winter coats or layers of fat, but by Thanksgiving a small bird may have twenty-five to thirty percent more plumage than it did in midsummer. It won't grow any more of the stiff primaries, but the increase in the soft, high-loft little feathers with lots of "processes" (short, air-trapping fibers growing from the main barbules) is substantial. When it's cold, a chickadee can fluff this down to the point where the bird is almost round and the fluff and feathers are actually thicker than the bird's body. Compare that with the loft of your heaviest down coat; if you or I were set up like that, with a down oversuit a foot or two thick completely encasing us, we could get along tolerably well on a long winter's night (though, speaking for myself, the perch would have to be pretty stout).

Most expressions of sympathy for birds, at least the ones that I've heard, focus on their unfeathered appendages: eyes, beak, and feet. The eyes aren't a big problem, however, being mounted almost completely inside the head (which, along with the inner part of the chest, constitutes the animal's "core," maintaining the most constant heat even when some less critical parts cool off). The eyes are not likely to be at great risk, being so near the furnace itself. The beak is more at risk but far less sensitive to cold. It does open into the mouth, though, and contain the nostrils, but it can be conveniently tucked under the feathers if the weather is too bad to stand otherwise.

The bird's feet elicit the most sympathy, probably because humans don't have beaks but we do have feet, and we can't imagine standing barefoot in the snow when it's twenty below zero. Unlike ours, though, bird's feet are usually almost fleshless. There are sufficient firm tendons to do the necessary flexing and bending, but practically none of the softer muscle tissue that encases our own legs and feet. Anything organic, including bird feet, can freeze if it gets cold enough, but in winter there is some physiological mechanism that reduces blood flow to the feet to the minimum needed for retaining at least some functioning ability. The hard tendons are pretty safe from freezing.

But these adaptations and comforts hardly seem enough to do the job for the chickadee, and as it turns out they aren't. Small birds have highly-geared systems. They burn off energy fast and must consume pretty steadily or run out of fuel. Unless they have found a way to save food (gray jays store it; redpolls have a pouch in their esophagus to hold extra, as well as having an unusual ability to feed in dim light) or decrease their need, they're never going to make it through a fifteen-hour night without feeding.

Decreasing need is the chickadee's evolutionary strategy, and unlike down coats and warm shelters it's one we can't well imitate yet. Hypothermia is a dreaded condition in humans, but it solves a big problem for chickadees.

A study done at Cornell in the 1960s showed that chickadees, like some other birds that have been tested over the years, can respond to extreme cold and repeated food deprivation (like what occurs every night in winter) by lowering their body temperature and thus reducing their various nutritional (food, water, oxygen) needs.

If it doesn't get too cold, a little shivering may be enough. Shivering, technically, is defined as "muscular contractions in an asynchronous pattern that does not result in gross movement of whole muscles." It is a way, in short, to generate heat through movement and sit still at the same time.

But after a certain point shivering is not enough. It ceases as the bird becomes torpid, entering some form of controlled hypothermia wherein body temperature can drop by a certain amount but no more (assuming that it doesn't become so incredibly cold that night that the bird simply can't recover to normal temperatures). In the chickadee, body temperature may drop as much as twelve degrees centigrade, which means that the bird's oxygen consumption is reduced as much as 75 percent. Bodily needs for nutrition are correspondingly reduced, and fifteen hours of fasting becomes possible. That amount of temperature decline in a human would give you a body temperature in the low seventies Fahrenheit, and your recovery would be a much more complicated process than the chickadee's.

We had a shocking cold spell last November. About the time it should have been getting pleasantly cold, maybe threatening a little snow, it suddenly became arctic, a cold almost violent in its surprise. It was ten below, sometimes twenty, for weeks, and the birds at our feeders, and at other feeders in town, just disappeared. We joked that they were all down at the Greyhound station buying tickets for Albuquerque, but in fact we were worried: the birds animate our backyard, and it was eerie looking out there day after day and seeing nary a pinion. We wondered if they'd all died, or moved away, or holed up, or found some better shelter with food nearby.

Chickadee Down

Finally, one day several weeks after the big freeze, I caught a flash of motion in the lilacs. Sure enough, there was a chickadee, and then another. For most of that winter they remained the only reliable visitors we had, and I don't know which I've enjoyed more about their survival, the mystery of how they did it this time, or the wonder that they did it at all.

Crawdad Eyes

If you hike, fish, trap, watch birds, hunt, or do anything else that requires spending much time along or in the shallow edges of streams and ponds, you will eventually come upon a crayfish. The first time you do, you will likely experience two realizations. The first thing you will realize is that the crayfish is already gone. By the time you notice that you have spooked a crayfish, it has made its getaway, perhaps in a little cloud of silt, perhaps under a rock, or perhaps to a new location, where it has settled and blended into the rocks and dirt around it. I have noticed that, though I occasionally see one take off, I have a lot of trouble telling where it went.

The second realization may not come for some time and may be in the form of a mental double take, as your mind fully registers what you saw and you begin to question it: "Did I just see a crayfish swim away from me *backwards*?" The answer is yes, you did; they swim backwards, especially when they are in a great hurry, as anyone would be whose daily rounds have suddenly been interrupted by an enormous foot.

When I first became aware of this peculiar trait, I spent a lot of time trying to figure it out. Why would any animal evolve to do such an odd and apparently inefficient thing as going as fast as it could hind-end-first through the water? Charmed as I was by the theoretical exercise of trying to solve this puzzle, I, predictably, didn't bother to learn anything about the crayfish first, so I never got very far in my inquiry.

But in a recent conversation with a trout biologist from New York, I happened to ask a question about geology. He smiled and said, "Well, since that's not my discipline and I don't know anything about it, I feel free to go ahead and theorize...." We both laughed, but I later realized he had perfectly described my approach to the mystery of the retreating crayfish. So I did some homework. It helped.

Crayfish aren't fish. They're crustaceans, members of the order Decapoda, ten-legged creatures closely related to lobsters, shrimp, and crabs. Crayfish look like small lobsters, and one of their common names is freshwater lobster; other common names are crawfish, stonecrab, mudcrab, and, probably most common of all, crawdad. There are other local names, some applied to only a single species. North America is home to more than 200 species and subspecies.

Our native crayfish are typically between three and six inches long. Most are muted in color, with that intriguing mixture of shades and tones that only nature seems able to produce, whether it be on the shell of a crayfish or on the bottom of a stream into which the crayfish blends so easily.

Some are more brightly colored. I once saw many bright-orange ones in the Siuslaw River near the Oregon coast, and they seemed so visible and vulnerable, sitting on rocks in the still, shallow water, that I couldn't imagine why such a brightly colored crayfish would ever have appeared on earth. All I could figure was that they had just molted, shedding their darker skin. Crayfish, like some insects, become a brighter color when they molt, before the new skin hardens and darkens. They also become less mobile, and slower—the gulls must take a terrible toll on them then.

Crayfish have one pair of long antennae, sometimes

nearly as long as their body, which can be directed all around for feeling and balance, and a shorter pair of "antennules," each with two flexible tips. Their compound eyes are mounted on short stalks that give them a wide range of vision.

The "deca" in decapod refers to the crayfish's ten legs, five on each side. These are highly specialized. The first pair are the large "lobster claws," known as chelipeds, which do most of the grasping and tearing of food (these can also deliver a mean pinch to the unwary naturalist). The second and third pairs have small claws useful for grasping food or footholds, but each leg of the fourth and fifth pairs has only a single nail. These legs are used mostly for walking. Crayfish are often missing one or more legs, which break off easily if the animal wishes to escape from a predator holding it by a leg. A new, full-sized leg will grow in over the course of the next few molts.

Crayfish routinely walk on their second through fifth pairs of legs. Some types can swim forward using the tiny appendages called pleopods, or swimmerets, that grow in pairs from the underside of the crayfish's abdomen (the abdomen is the part of the lobster you get when you order lobster tail). When speed is important, crayfish launch themselves backwards with a quick, downward-bending stroke of their abdomen, using their flat, fanlike tail to propel themselves along. Some species even do this just to get around, danger or not. This swimming motion is not unlike that employed by shrimp, who also bend at the middle to propel themselves through the water.

Crayfish are omnivores, and for all their intimidating armament, they are primarily scavengers and vegetarians. Most of their time is spent clambering about in the rubble and mud of the bottom in search of a meal. They feed mostly at night, and will sometimes even walk ashore for a look around.

They nest in burrows they dig in the soft bottom of a stream or pond, tunneling as much as a yard down. (In some species the female does all the digging, preparing the nest for the arrival of a male, with whom she will mate.) Soil from the tunnel is formed into small pellets that the crayfish carries back to the entrance, where it builds a mound, known informally as a "chimney," as much as a foot high around the entrance.

The mating behavior of crayfish would give even the least anthropomorphic naturalist pause to wonder at how the little crustaceans ever got so common; the procedure is dreary almost beyond human imagining. One ecologist described it as "more or less a matter of chance. The male has no power of sex discrimination, and during the mating season he seizes and turns over every crayfish coming his way. Another male will always resist strongly, but a female will either resist or remain passive and receptive."

Mating season varies by the species, ranging through the warmer half of the year, when the crayfish are most active. In some species, the male's sperm are held separate from the eggs in a cavity in the female's body for weeks or even months, then exposed to the eggs when the latter are ready. Females carrying clusters of the globular eggs on the underside of their abdomen are said to be "in berry." When the young hatch, they usually cling to the mother for a couple of weeks before going out on their own. When fully grown, the males of a typical species will have larger claws and a thinner abdomen than the females.

Armed with this much information on the crayfish, I find its propensity for reverse gear much less mysterious. It is a member of an order of animals who, though they live with fish, have lots of legs. Think of the crustaceans you've seen: crabs scuttling sideways across rocky beaches, lobsters jostling each other in restaurant tanks. They walk in any direction they choose, often without bothering to point their faces that way first. (Part of the creepiness associated with watching multilegged animals walk must be that they do it so differently from us, making us uneasy with their disregard for mammalian conventions such as aiming your head toward where you want to go before you try to go there.)

Having legs in water is different from having legs in the air. On land, legs can allow you to escape from an enemy; in water, as anybody who has tried it knows, running is a lot less efficient than swimming. The crayfish's evolution was directed in part by the reality that whatever good its legs might do it, they wouldn't carry it away from predators. We shouldn't be surprised to find that the crayfish doesn't turn around and run away.

Besides, it is a forager, and a tunneler, and a tearer. It has developed a set of appendages—claws, antennae, feet— that allow it to do those things very well. But an animal can only succeed at so many things at once. The crayfish can climb, grab, dig in, and handle all manner of things in front of or beneath it, but those stiffly jointed legs do not have the range of motion to be good oars, especially for high-speed competitions.

A crayfish on its feet reminds me a little bit of a backhoe: even if, after considerable maneuvering and scuttling, you get it turned around to go a new direction, it's still a backhoe and won't go all that fast. But if the crayfish can keep all its main fighting and digging tools facing an enemy, then perhaps by backing that flattened tail and abdomen under a rock until only the big claws are exposed, it won't have to worry about fleeing.

Crayfish make their livings in the quieter waters of streams and in ponds, where they are constantly in danger of attack. I doubt that any crayfish you scare up didn't know before you appeared just what direction it would head in if you did; they must live in the habit of knowing where the nearest shelter is. And I doubt that when they do swim, they ever swim far. For the crayfish, as for a good base-stealer, long-term speed is nowhere near as important as instant acceleration and a good sense of what's going on around you.

That's where the crayfish's eyes make such a difference. They're mounted on the top of the head, on little stalks, and they give the crayfish a great bowl of vision above, beside, and even behind it. A crayfish moving backwards doesn't suffer from the same problems I would if I tried to run backwards. It doesn't have to turn its head (even if it could). With those top-mounted, stalked eyes, it can no doubt see more than enough to go where it wants.

In the process of sorting all this out for myself, I asked a friend of mine, an ecologist who specializes in freshwater life, what he made of the crayfish's backward swimming. He suggested many things, but what stuck was his criticism of me for seeing backward flight as a problem in the first place. The notions of backward and forward are human concepts, and they're rather loaded terms; something that runs backwards, the way comic movies do, seems silly to us.

But what does it matter to a crayfish that his face is aimed in one direction and he is headed in another? He can still see where he's going, and even if he couldn't, he'd at least be moving away from danger and would be able to keep good track of that danger because he was looking right at it. In open water, especially if the crayfish has no refuge in sight, the most important piece of information at any given time is the location of the danger. What better way to keep track of it than by swimming backwards so you can always watch your pursuer?

Which brings me to one last question, admittedly an emotional and very human one. What must it be like? Imagine that as you dart this way and that, you have to watch a living nightmare approach—let's say it's a big bass or trout—surging through the water behind you, appearing and disappearing in the silt you stir up, its mouth open as it strains to engulf you. I don't suppose that the crayfish is able to give that any thought, but I'm just as glad to be an animal that can avoid knowing too much about what's after me, by turning around and running like hell.

Bird of Paradox

It's funny where you find your friends. I first went trout fishing for reasons I'm sure I didn't understand, but that had a lot to do with just being out there. I appreciated the old saying, even if I couldn't have articulated it at the time, that fishing has to do with a lot more than just catching fish. But I had no idea of all the places it would lead me. Some of them I've written about before, and some of them I write about here and there in this book. It's still leading me places, and one of the great pleasures of anticipating those new places is knowing that I will have wonderful company in the form of a plump gray bird named the dipper.

The dipper, a robin-sized bird long known as the water ouzel, is a common resident of many western streams. It stalks the shallows on its spindly songbird feet, and it dives eagerly into the faster water, seeking the aquatic insects and tiny fish that make up its diet. But it didn't get its name from taking so many dips in the river. It has the engaging habit, exercised every few seconds, of dipping its entire body down in a little knee bend, then rising back up to its full height. It does this quickly, and it does it all day long.

If Disney were to portray the dipper in a cartoon, the

bird would probably become a fussbudget, a bean counter among birds. Its almost uniform slate gray plumage (the head is sometimes a little rustier, and many feathers sometimes have the faintest end-rim of white) makes a perfect gray flannel suit, and the quick hopping, bowing gait lends itself perfectly to a high-pitched, anxious little bureaucratic voice—"Oh my goodness gracious [dip], so much to do! Oh [dip], good morning Mr. Mallard, how are you today? No time to talk [dip], must be about my business. Yes, yes, so much [dip] to be done and already it's [dip] lunchtime!" I almost hope the dipper remains anonymous, and regional, enough to avoid such attention. It deserves better. I rank it with the chickadee, the phoenix, and Tolkien's great war eagles of Middle Earth in my personal gallery of birds that hold my attention beyond all reason.

Here is a bird of boundaries, an animal that has evolved to fill a niche that seems not to be a niche. In every respect I can think of, the dipper is a portrait of compromises that allow it to live in two worlds.

The most obvious boundary is between air and water. The dipper flies in the air and, in an oddly altered way, also flies in the water. In both cases, the wings serve to propel the bird forward. In the air, they lift it; under water they help to hold it down, to keep it from bobbing to the surface. The dipper makes the transition without hesitation, even plunging from the air directly into the stream and shifting its propulsion gears into effortless swimming, then bursting from aquatic flight back into aerial flight.

Another boundary is physiological, between the typical water birds and the typical songbirds. The dipper is consistently typical of nothing except dippers. It is a true swimmer and underwater forager, like ducks, yet it is a competent (if infrequent) percher, with no webs on its feet and no shovel or great extension to its beak. It is shaped like an overstuffed wren, hardly a likely form for a bird that must move through powerful, swift currents, but it has a preening gland a thousand percent bigger than other songbirds its size. My overpriced, high-tech raincoat should be so waterproof.

Another boundary is between habitats, and few birds define their range with such fidelity to a single feature, in this case the surface of their chosen stream. I can't think of any bird that approaches the dipper for single-minded devotion to a trail of water. Many observers have pointed out that the dipper almost never flies over land, at least not more than a few yards from the river's shore; that the dipper prefers to fly only two or three feet above the water (which makes it seem faster than it probably is) and will typically fly over you, as you stand in the stream, rather than around you, if going around means it must cross land. Rather than make a beeline across land, it will faithfully follow the most meandering of streams, flying only a few feet above the water, for a great distance, and only turn from the main current when it can follow a smaller tributary. Even its courtship flights occur over the water. Observers also note that it will sing while so close to noisy cascades that no other animal more than a few feet away could expect to hear it; that its nests are built so close to the water that they are frequently kept damp by the spray (some early naturalists supposed the dipper did this on purpose, so that its nesting materials would stay green and thus less obvious to predators); and that the young require virtually no training to handle themselves adequately in the water. At times the dipper seems more a behavioral prisoner of its own evolution than an animal that has found a unique and utterly satisfactory home range. Ducks, sandpipers, killdeers, herons, and all its other waterbird neighbors fly high and far from the water; the dipper stays home.

And it stays home all year. From the earliest days of serious American bird literature, naturalists have expressed admiration bordering on awe for the dipper as, in the words of one naturalist, "the very quintessence of hardihood." Even in milder winters here in the northern Rockies, the headwaters of many streams freeze over, but the dipper only moves far enough downstream to find open water.

On my own home river, which rises at 9,000 feet and whose lowest reaches are still a mile high, dippers are common here and there all summer, but in the winter the cold pushes them down to the warmer elevations. Then they stack up along the shore so that there's another one or two every hundred yards or so. I often walk the lower river in winter, and it is then, more than any other season, that I know the wisdom of the mountaineer/conservationist John Muir's grateful tribute to this somberly dressed little bird:

> Among all the mountain birds, none has so cheered me so much in my lonely wanderings—none so unfailingly. For winter and summer he sings, independent alike of sunshine and love; requiring no other inspiration than the stream on which he dwells. While water sings, so must he; in heat or cold, calm or storm, ever attuning his voice in sure accord; low in the drouth of summer and drouth of winter, but never silent.

This is no less true for being so affectionate. The dipper is an almost indefatigable singer, keeping up its bright and surprisingly loud song when the surrounding woods are silent. In fact, were they to abandon my stream there is only one thing I would miss more about dippers than their song.

It is going on twenty years since I first shared a river with a dipper, and even yet there is nothing that can so instantly distract me, transform my mood, or generate a nearly daffy grin as the bobbing of a dipper. There is in that little performance an energy, an industry, and all indications of an irrepressible spirit that will not let go of my attention. It has always brought me somewhere closer to mirth than I was before I saw it, and I tend to get pretty mirthful along trout streams anyway. It means that much to me, even though it is still unclear just what it means to the dipper.

Dipping is a pretty little mystery. Received wisdom has it that the dipper developed its peculiar little "nod and curtsy" (as Muir described it) to compensate for spending so much time near streams, where its song could not be heard. It is more curtsy than nod; the whole bird dips as its "knees" lower it into a crouch, then it springs straight up. This, according to one theory, alerts its companions to its presence, regularly, where the stream is making too much noise for more traditional bird-song communication to work. Some have said the dipping occurs as much as once a second, but I don't see it happening that frequently. On the other hand, one naturalist reported (in 1936) that a dipper remained motionless for eight minutes—behavior so uncharacteristic of these nimble birds that I have trouble believing it. I've never seen one go even thirty seconds without a dip.

Those who claim that the dipping is a form of communication cite as proof their observation that the motion becomes more frequent when the bird has been disturbed. Others aren't convinced of the correlation. One study of wild birds captured and placed in a variety of contained environments recorded that dipping nearly ceased in captivity—and what could be more disturbing than captivity? Just the other day I approached one bird to within about fifteen feet, and it looked to me that when he noticed me, he froze for ten or fifteen seconds (a long time for a dipper) before resuming his dipping.

Some naturalists have said that older (and, by implication, wiser) birds dip more often than do young ones. Others say that, though most don't dip at all until they are at least a couple of weeks old, after that it's hard to tell who dips most.

I can't imagine that dipping is just something the bird's genes brought along for no good reason from some ancestral Original Wren; it's obviously an important part of the bird's

activities, and it makes sense for a bird, even as vigorous a
singer as the dipper (whose loudness may also be a response
to the proximity of streams), to have some alternative way of
keeping in touch with friends.

It's obvious as well that the bird needs such an alternative. The first rule of dipper-watching is: Try not to look away.
The bird so perfectly matches the muted, mottled grays of the
wet, damp, and dry rocks of the waterside that even the time
it takes to raise binoculars to your eyes and refocus your vision through them is time enough for the bird to blend out
of sight for a moment. Your best bet then is to lower the binoculars and scan the area where you last saw the bird, hoping
its next dip will reveal it again. Even assuming, as I do, that
other dippers are much better at keeping track of dippers than
I am, the curtsies must help them, too.

On the other hand, most of the dipper's reported predators are at its level: right along the streamside. Its color and
its totally vertical (rather than lateral) bobbing probably help
keep it invisible to flying predators, but the minks, snakes,
skunks, and other carnivores who meet it at eye level must
take some advantage of the dip. This apparent shortcoming
in its behavior may mean nothing (it could be explained by
Paul's first rule of survival adaptations: You can't do everything right at once), but at least it suggests that the dipping
serves purposes so important that the risks are not unac-
ceptably high.

An artist I know, someone who has drawn many animals from life, pointed out to me that of all the animals she
had studied, humans are the only ones who normally show
much of the white of their eyeball. She wanted me to tell her
why this was. I had no idea, but I've wondered a lot about it.
When we see another animal with white showing, we are
alarmed—it looks "wild-eyed." A horse breeder recently told
me that Appaloosas, which he raises, are thought of as a little

goofy because they tend to show more white than other horses. When any animal, human or other, bulges its eyes at us, we tend to react with fear or at least concern. Eyes are extremely expressive communicators.

When I watch dippers through binoculars, and bring them in close enough for a really good look, I find myself having some pretty complicated reactions to their eyes. Like most birds, dippers have a "third eyelid," known as a nictitating membrane (see more about this in the "Microdragons" chapter). This is a sliding membrane under the eyelids that apparently serves the purpose of cleaning the eye, an especially important service for a bird that spends so much of its time facing into fast, sediment-stirring water. The membrane is opaque but appears white from any distance.

The dipper's eyelids also are either white or white-edged, so that whether it blinks them or the nictitating membrane, or both, there is a sudden white cover over the eyes. Against the dull dark background of the often wet feathers, and in certain light, these blinks are unbelievably bright, almost like little flashes of strobe lights. When I first saw the flashes, my mind hardly wanted to register them, and I've never gotten used to them.

The effect of the flashing eyes is strangeness; what at first appears to be just an odd little bird with a hopeless nervous tic becomes something nearly bizarre. It's easy to imagine that the animal is mechanical, and its eyes are the only giveaway of the inner works. Its eyes flash at you, and you feel photographed, or exposed, or understood a little better than you'd like a bird to understand you. Disney could make nothing demeaning of this; it is too arresting to be trivialized.

So I watch the dippers. I feel the affection for them that Muir expressed so well, and I sometimes laugh aloud at their bouncy little routine. They are always good company.

But mostly I see them from a distance as they poke

around in the shallows, or fly with their steady, buzzing wingbeat upstream or down. Under those circumstances, I don't have to deal with those eyes.

Now and then, however, and without especially intending to, I look for the eye flash that is almost out of range of my vision. Sometimes I half-imagine I can see it, but often I'm not sure. Most days when I do, it seems an innocent enough little wink, just an accident of physiology. But other days it has that alien, almost sinister quality about it, and there along my home river, the place where I feel more comfortable and assured than in almost any other, I wonder if the dipper is really my friend.

Antlers Aweigh

A few years ago I spent part of a wet spring fishing the Au Sable River in Michigan. The river was in flood and the fishing was poor, but even then I had grasped a wisdom I recently saw articulated in a bumper sticker, that "poor fishing is better than any kind of work." It had gotten dark, and I was about to wade out and go home, having also grasped the wisdom that poor fishing in the dark in a flooded river is better than any other kind of suicide. I had just made my tenth "last cast of the night" when I noticed something bobbing along in midcurrent upstream, at the next bend.

At first I took it for a chunk of tree, the sort of hazardous flotsam that appears in rivers any time there is high water, but it soon resolved itself into the head and upper neck of an antlerless deer. She was swimming along with the current, the rhythmic pulling of her legs evident in the rise and fall of her head. Standing in the shallows with solid forest behind me, I was invisible to the deer, so I watched her until she was almost even with my position.

As she approached, my curiosity got the best of me and I had to know how much control she had. I slapped the heavy fly rod down on the water in front of me a couple of times, startling her. She made a fast right turn and with surprising disregard for the powerful current swam straight to the opposite bank. Once out of the water she paused for a moment, looked back toward the source of the sound, then disappeared into the trees.

My first thought was that something—probably a dog—must have chased her into the river. But then, thinking over her obvious familiarity with water, it occurred to me that she was simply going somewhere by the quickest route. I was the only interruption in her evening.

Deer are confident swimmers. Most of what exists on the subject is scattered here and there in sporting literature, that immense, terrible-yet-wonderful body of writing that until the last few decades was often the best published source of natural history information we had. Sportsmen in the nineteenth century frequently wrote about deer swimming across lakes, and one of the many debates of that time among hunters was over the sporting ethics (or lack thereof) of killing a swimming deer. By 1900, it was pretty much agreed that only a lout or a moron would shoot a swimming deer, or use the other common approach of paddling alongside in a canoe and cutting the deer's throat as it swam.

There were also nonhunting accounts in the periodicals. *Forest and Stream* (out of circulation for more than fifty years now, but still my favorite outdoor magazine) reported in 1896 that a deer had been sighted more than a mile off the Rhode Island coast, industriously swimming out to sea. Others have been seen as much as five miles from shore in both the Atlantic and Pacific oceans. Ernest Thompson Seton, the great early-twentieth-century nature writer, gathered up several such accounts and concluded that deer are "so confident of their swimming powers that they invariably make for the water when hunted to extremity. There are many many cases on record of deer so pushed, boldly striking out into the open sea, trusting to luck for finding another shore." The famous tiny deer of the Florida Keys are known to swim from island to island, movement apparently important to that strain's genetic well-being because it allows the small populations on each key to mix now and then. Early one winter morning on No-Name Key, I saw one cross the road in front of me. It was no larger than a small dog. The idea of such a small animal swimming the deep open channels between the keys makes me wonder about sharks, for whom such dainty little deer would be a handy meal, and thinking about sharks naturally makes me think about fishing. It was fishing, after all, that got me wondering about all of this in the first place, and it was fishing that taught me just how interesting the swimming of deer is.

The interest for fishermen lies in deer hair. Deer hair is among the most popular materials for making lures, especially flies. Native Americans used deer hair to fashion a variety of fishing lures. Some would troll an entire white-tailed deer tail with hooks in it behind a canoe to catch muskies, pike, and other large freshwater fish.

The great naturalist William Bartram, on a trip through the South in 1774, saw Florida Indians fishing with a "bob," that being a large treble hook around which was lashed part of a deer tail, red cloth, and colored feathers, a bundle "nearly as large as a fist." The bob was then tied to a line about two feet long on the end of a twelve-foot pole. As one man quietly paddled the canoe along a lakeshore, another stood in the bow and waved the bob around over the surface of the water near weed beds. The explosive strike of a fifteen-pound largemouth bass as it rushed from the cover of the lily pads and weeds to attack the bob must have been as good sport as any had by the most overtackled modern angler.

The modern "bucktail" flies resulted from the availability and usefulness of deer tails, whose long, fine hairs make excellent, lively imitations of small baitfish. But bucktail is not the deer hair of greatest interest to me, either as a fisherman or as a naturalist. The body hair is, because it has so much to do with the deer's ability to swim so well. Deer body hair contains countless air cells (it is sometimes described as "hollow," but it doesn't have an empty core) and so is wonderfully buoyant.

An adult deer has several kinds of hair, but for my purposes it is the common long hair, known as guard hair, that matters. Fly tiers use this tough, flexible hair in a remarkable variety of ways, packing it densely on a large hook and trimming it into the shape of a frog or a mouse, or tying just a few pieces on a dainty hook to imitate the outline of a mayfly's wing. The stuff is almost unsinkable.

Deer molt twice a year, wearing a light summer coat for five months and a heavier winter coat the rest of the time. The old hunters understood this; the sports who shot swimming deer knew that if the animal was in its "red" summer coat someone better have a good grip on the tail before the animal was shot, because it would sink as soon as it stopped swimming. If shot in its "blue" winter coat, on the other hand, it would probably float for quite a while. The winter coat, being thicker, held more air, though the deer's layer of winter fat also helped flotation.

That doe I saw in the river was probably almost shed of her winter coat; swimming was serious work for her, work that apparently had more purpose than I realized at the time. All my readings about deer in water described cases of deer traveling across it just to get to the other side, or entering it to escape some pursuer. My deer was doing a little more than that; she was taking a ride on the current, for whatever reason. She seemed to have learned that she could travel quickly that way, though of course I have no idea if she knew where she was going or if she was just in the mood to go. Perhaps she used the river routinely, to travel from one part of her range to another.

I would trade all the painstakingly made little deer-hair flies in my fishing vest for another chance to watch her go by. This time I wouldn't spook her. I'd let her pass, and then, if I could walk the shore as fast as the deer was swimming, I'd follow. With a little luck maybe I could stumble along in the dusk quickly enough to learn her destination.

The Longest Meadow

A friend of mine lives in a little cabin about half a mile from the nearest public road. She keeps a horse there, sometimes in a small corral and sometimes just letting him run loose in the nearby hills. This is in the Rockies, so when the sun sets on a summer evening even a horse can get cold. He has the engaging and troublesome habit of drifting down the valley and settling his belly on the sun-warmed pavement of the road, much to the surprise and consternation of motorists.

For wildlife, roads are rarely benign. A highway is not always harmful, but it is always an important presence in the life of the natural community it intersects.

The influence of the road starts with the pavement itself, and may be especially noticeable in dry country. On a recent trip through eastern Oregon, I was struck by what the pavement does to the roadside vegetation. First, the construction of the road creates a narrow band of disturbed soil that is ripe for all sorts of plant colonists. Then, because of its nicely beveled surface, the road drains whatever rain falls into the disturbed soil.

In a dry climate this can make a huge difference in the

soil's hospitality to plants. That slight amount of moisture, rather than being spread out over the entire area where the road didn't used to be, is concentrated in a narrow band of soil right along the pavement. It is the equivalent of increasing the precipitation in that soil by several hundred percent, and it is probably the biggest reason that in the Oregon desert that August, while the nearby land was already parched, there were tall, bright yellow wildflowers blooming along the road like someone had just transplanted them there that morning.

I suspect a similar but less pronounced effect is had by the ditch that invariably accompanies the road on both sides; water drains not only from the road but from the adjoining lands into it, and if the ditch is deep enough, the soil in the bottom may get enough less sun to be more moist and productive. I also suspect that the intense heat generated by the blacktop has some important effects on the soil and moisture underneath and near it, but even without those effects, the road makes a big difference in how nature behaves. Most of that difference is apparent in the activities of wildlife.

I suppose that for most people what comes to mind at the mention of roads and wildlife is deer. The destruction of deer by cars is extraordinary: by the mid-1970s Pennsylvania was reporting over 20,000 collisions a year, and Michigan was averaging 16,000. That is not only a substantial public safety hazard, it is a sizable population sink for the deer. Most other wild animals, from the smallest snake to the largest moose, suffer proportionately. Estimates run as high as a million a year.

The relationship between roads and wildlife has been of interest to ecologists for more than sixty years, almost as long as the family car has been able to go fast enough to do animals harm. To the casual observer it appears only that roads are a long, narrow slaughterhouse, but ecologists have long known that roads have their advantages, too. Stereotypically, human influence on a natural setting is assumed to be bad, something that destroys balances and reduces diversity by eliminating fragile life forms. Country roads are a superb example of how human influence can increase, rather than reduce, the elegant complexity of a natural setting.

The road is not simply something that animals must get across now and then. It immediately becomes a part of the habitat, to be used whenever possible. Road-killed animals attract other animals (imagine the protein in 20,000 dead deer), so that a dead deer feeds an assortment of bird, mammal, and insect scavengers who themselves are then made more vulnerable to passing cars.

Road construction creates thousands of miles of very narrow meadow interspersed with thickets and lined with power lines and fencerows, so that songbird losses to cars are in some cases outweighed by improvements in habitat for more birds. Hawks and owls hunt near roads, taking advantage of all the new meadows and perches, and learning that small animals are especially vulnerable and exposed when they cross the road; in winter, the road might be the only such exposed spot for miles. Of course, every new animal that shows interest in the road becomes potential fodder for the car-mill (is that why they call it *carnage?*), and animal protein moves through the natural system steadily. I used to live in southern Vermont and was in the habit of bicycling along Route 7 just north of Manchester. On no day did I see very many dead birds along the road, but rarely on two different days did I see the same bird—someone is always looking for the free roadside lunch.

Animals adjust to unusual opportunities. I've seen crows develop almost unbelievable road savvy, so that as my car approached them when they were feeding on something in my lane they waited until the last possible instant and then casually hopped one step over the median line to let me pass.

I have noticed that ground squirrels here in the West would sometimes gather around a dead relative, and at first I wondered if these little vegetarians were changing their diet. Finally a sharper-eyed friend noticed that they weren't eating the dead squirrel, they were eating the undigested contents of its stomach. There is a lot of creative adaptation going on here.

And there is a lot to be adapted *to*. The road's influences kind of dribble off into nearby areas. More than fifty years ago no less a wildlife authority than Aldo Leopold noted that spreading salt on and near roads (back then it was done to keep down dust) attracted deer to heavy traffic areas. Even moose, according to a recent Canadian study, are drawn to salt pools that form along roads when melting snow and salt mix and drain away. But salt is only one, easy-to-see, factor. Several researchers have recently reported that lead contamination, especially in areas of heavy traffic, builds up in roadside soils and then works its way up the food chain through vegetation, insects, and mammals. This process adds a subtle element of treachery to the road's banquet invitation.

My own interests in this situation involve the smallest road-killed animals, insects. I used to wonder, as I drove along to a symphony of splats and thuds, if the insects were as thick over the nearby pastures as they were over the road. They may not have been. I suspect that the heat the road gives off, which may be the insect equivalent of a thermal, has something to do with dense insect presence, but a wise old fisherman has suggested another good reason. Aquatic insects—mayflies, stoneflies, caddisflies, and so on—that emerge as adults from water and then fly off to mate, often mistake road surfaces for their riverine homes. Anywhere a stream or pond is crossed or parallelled by a road, a certain portion of the mating flight may get confused. I've watched large stoneflies, fluttering upstream on their way to laying eggs in the water, take a wrong turn as they fly over a bridge and head up the road to inevitable disappointment when they try to dip their egg-laden tails in the hard pavement.

Next time you're driving in the evening, when such emergences are most common, watch carefully as you cross a bridge or pass a pond; the windshield will tell all. The tendency of bugs to mistake the road surface for something else may go a long way toward explaining the invention of the squeegee. My own informal studies, conducted under all imaginable driving conditions over the past twenty years in forty-eight states and all Central American countries in existence before last Friday, indicate that the daily mortality of airborne insects over rural roads is typically on the order of a zillion per mile. Most have just finished dinner.

There are millions of acres of public land along roadsides, and most of it could be or is managed to enhance some kind of wildlife habitat. Wildlife managers and biologists, being opportunists, have discovered that roadsides are not only good for wildlife, they are good for learning. Techniques have been developed for taking an accurate census of raptors from road surveys, or to learn various population characteristics of deer from studies of road-killed animals. In many cases the only chance biologists have to examine large numbers of deer or other animals is during the hunting season, and that has some serious limitations: it provides data only for a short season, it may apply only to males if the species' females are not hunted, and it only provides data on game animals. Studies of road kills can be conducted year-round, which may eventually yield, for example, data on deer does throughout their pregnancy cycle, or on songbird abundance through migration seasons. A study of the stomach contents of road-killed coyotes may say a great deal about local rabbit population trends. For those of us who would rather read *The Journal of Wildlife Management* than *People* magazine, this is all great fun.

A few years ago, a friend and I were bicycling on a deserted road in southern Florida. There is a surprising chill after sundown in the Everglades, and alligators frequently come up onto the road, like my friend's horse in Wyoming, to warm their bellies on the pavement. At the same time that we were watching for alligators, we were envying them that nice warm road.

Finally, still having several miles to go, we decided we just had to give it a try. As we stretched out on the pavement, giving front and back repeated bakings, I gained a new respect for horses and alligators. The road was almost as warm as it had been at midday, and I suspect it stayed that way through much of the night, long after we got back on our bikes and pedaled home.

Nellie's Dream

Very early last spring here in the Rockies, I was out with a couple of friends, grizzly bear researchers. There was still lots of snow, and we backtracked a mating pair of grizzly bears a short distance to their most recent daybeds. They had left the daybeds a couple of hours before but were obviously nearby. Their tracks and droppings told an interesting story.

As they had left the daybeds, which were just flat depressions under two neighboring trees, they had moved uphill. Three times within the first hundred yards, the female stopped to defecate. Each time, the male (who was huge) stopped and with uncanny accuracy dropped a load directly on top of hers.

None of us had heard of this. It seemed most likely that the male was making some kind of territorial statement—She's mine, don't get any ideas—to any other bear that might come along and investigate the droppings. There was probably more to it than that, but that was enough. At the time I joked that his technique was quite effective—I had no desire whatsoever to offer to buy that female bear a drink.

But because the droppings were so large, we began

wondering about other things. For example, what if you had never seen or heard of a grizzly bear? What if you were walking through the woods and you suddenly came upon those fresh piles, or any such pile produced by the big male alone? You would instantly know what the stuff was, but what sort of creature would your imagination build around the orifice required to produce it? Your data base is small, but it gives you lots of room to wonder.

What our imagination would make of bear scat is an appealing and amusing question, but it leads me to harder and ultimately more enriching questions. If a grizzly bear has occasionally starred in some disturbing dream of mine, what sort of god-awful thing chases a grizzly bear through *its* dreams?

That question invariably leads me to thoughts of Nellie, a Cardigan Welsh corgi of whom I once enjoyed co-ownership. In case you've never seen one, corgis are long, low-slung herding dogs, apparently one of the oldest breeds of working dogs. A big male might weigh thirty pounds, but they're known among their admirers as "the big dog in the little dog's body" because they are so tough, fearless, and hardworking.

Anyway, Nellie had dreams too. At least I must assume they were dreams; as with so many other dogs I've seen or heard of, her feet would move slightly, her breathing would quicken, and she would utter the softest little whimpers of excitement.

The dreams of another species of animal provide an unparalleled opportunity for wonder. We wish we could see them, somehow televised on a screen, perhaps, and suddenly enter a fantasy world of almost imponderable difference from the wildest imaginings of our own.

With dogs, of course, it might turn out that some of the dreams are relatively dull visually—table legs, muddy car tires, puzzling glimpses up heavy overcoats. What we would really need to appreciate a good canid dream would be something like the full-sensory "feelies" in *Brave New World,* theaters where all the senses are involved. What a richness of experience the dog dream might take on if we were exposed to it with the dog's capacity for picking up, cataloging, and enjoying scents, and the dog's attuned hearing, which recognizes its master's car two blocks away on a busy street just by some peculiarity of the tire tread hiss. Sometimes, as Nellie walked across the living room, she would encounter an interesting scent on the carpet. She'd press her nose against it and with a deep inhalation drink in the whole message. You could just see the computer screen race along as the data piled up.

A friend of mine had a very expensive hunting dog named Nimrod. Nimrod was the village idiot of the family's several superb hunting dogs, and that may have been why he got so much love at the same time that he got so much ridicule. Nimrod caused a bad wreck one day out in front of the house, by standing in the road over a dead skunk, sniffing at it just as a car came around the bend. The car swerved to miss Nimrod, and hit a tree. What struck my friend about the whole thing was how amazing a dog's nose is. He said something like, "Nimrod was there with his nose right up the skunk's anus! You'd think that with such a sensitive nose, a dog's whole system would be burned out by such an overdose!" Add that to your idea of the dog's dream: not only can it smell *everything,* it can stand the smell of *anything.* What a dream it could have.

But I was telling you about Nellie. Here she was, a perfect product of more than a thousand years of breeding and experience, an animal totally directed by its masters and by its extraordinary genetic plasticity to herd cattle and sheep. Nellie very nearly existed solely for the glory of doing that one thing well. The urge to do it must have been the closest thing to instinct that humans will ever introduce into an animal. Yet she had never seen a sheep or a cow.

And so, while I wonder what vast, horrific thing might

haunt the dreams of a grizzly bear, even more I wonder what vague, heavy-footed specter fled Nellie's nimble rushes across the mists and moors of her ancestral pastures. I wonder if she had some genetic memory that kicked in, to fill in the gaps in what all this should sound, and smell, and feel like. But as much as anything else, I wonder how much of the satisfaction—of dodging clumsy ghost hooves, of cutting off the specter's escape when it broke, and of bringing it to where she somehow knew it should go—she kept with her when she awoke.

Part II

WILD SOCIETY

JIM HAYS

Keep Moving

During a period of my life when I was writing full-time, I discovered that one of the delights of working at home is the coffee break. As a truly creative coffee-breaker, I could turn what should have been a momentary distraction into high art, puttering around the kitchen, checking to see if the refrigerator had hatched any surprises during the night, and invariably drifting over to the kitchen table, where I would sit down and watch the bird feeders in the backyard.

There are few things more helpful to a determined time-waster than the comings and goings of birds at a feeder. Feeders, with all their fuss and bustle, offer the casual observer very nearly the same vicarious sense of accomplishment as must be felt by sidewalk superintendents at a construction site: I usually had a vague and comforting feeling that I was in fact engaged in a useful, even productive activity (perhaps I was right—here I am writing about it).

It was the intensity of the distraction that made feeder watching so satisfying; whatever I was working on at the time, however absorbing it may have been, faded from my mind instantly at the sight of the feeder, replaced by a crowd of old, well-loved questions: Why do the finches settle in and eat seed after seed from the feeder, while the chickadees fly in, take one seed, fly to a branch on the lilac bush, eat the seed, and then repeat the whole process? Why does the grackle knock dozens of seeds to the ground for every one it eats? Why do juncos feed individually (and only on the ground below the feeder), while the siskins feed in flocks (both on the ground and in the feeder itself)? Why does a flock of goldfinches settle on the thistle feeder for a few minutes, chow down like famished dockworkers, then roar off to parts unknown when there's still plenty of seed left?

It was the last question that finally one day turned a coffee break into work, and sent me searching the literature of bird behavior for some answers. It seemed silly, and less than efficient, for a wild animal to abandon a sure thing—a perfectly easy high-energy food source—and go looking for something else.

Usually the first thing the nonscientist learns when pursuing a question like this is that scientists have not only been asking the question for a long time, but are asking (and answering) it in amazingly complex ways. How animals feed is addressed in a vast literature under the general heading of "Foraging Theory," the subject of numerous books (which, the newcomer almost immediately notices, start by disagreeing with each other). Just how those chickadees and siskins get through the day is analyzed and summarized in terms of such things as "multistage decision making," "maximizing energy gains subject to time-budget constraints," "exclusivity of search and handling," and "risk-sensitive foraging." These discussions are accompanied by charts, figures, graphs, and occasional formulas composed mostly of symbols that don't appear on any of my typewriter keys. This level of sophistication, whether I can penetrate the text or not, almost always serves to make me think hard, if only in imitation of the *real* thinkers.

I start by peeling away the fellow-feeling that has gotten

in my way. Because of their dietary preferences and their metabolisms, birds don't eat like people, so I shouldn't assume that their nutritional needs can be satisfied by the same means ours can. Your average American human will sit down for three short bouts of feeding a day, with occasional recreational consumption in between. A bird, like most other plant eaters, devotes a much larger portion of its waking moments to feeding. And a bird, unlike a human, doesn't find its food at one central spot with a tablecloth.

Take the chickadee. Its daily feeding routine is a tour of what ecologists know as "patches," separate little concentrations of food. There may or may not be bird feeders among those patches. The bird, rather than just wandering around the neighborhood looking for eats, is in all probability checking out a series of known (that is, previously visited) patches at the same time that it keeps an eye out for new ones. Movement is dictated by countless decisions: how long to stay at one patch, how much to use a patch that exposes the bird to greater danger than other patches but is richer in food, how heavy a competition from other birds to tolerate before abandoning a patch, and so on.

It is a world of trade-offs. Chickadees need a perch in order to feed; they hold the food between their feet as they eat it. Thus, they cannot feed at feeders that have flat platforms or large, thick perches that they cannot grasp. This means that the chickadee is bound to do a lot of back-and-forth between cover and food. Even if the feeder is equipped with usable perches, the bird may return to cover with each seed if the cover provides sufficient protection from predators or weather to justify the energy consumed in the flights back and forth (one chickadee in a 1984 New York study carried two seeds to cover each time, eating them one at a time).

A Cassin's finch, on the other hand, can sit at a thistle feeder and pick the little thistle seeds out till the cows come home; for the finch, there is apparently sufficient profit (that is, improved odds of survival) in that rapid continuous feeding to justify the risk of exposure to predators. The finches at my feeders almost always appeared in groups, and the combined vigilance of the group probably also contributed to the equation by reducing the risk of predation.

Even when the chickadees appeared in groups, most eating was done in the lilac bush. I noticed what I later had confirmed in my scientific readings, that any such flock or group had a clear dominance structure that influenced who fed when and how often. The most dominant bird had the power to take the least chances; the least dominant bird had to take higher risks to get the same amount of food.

All of this leads away from the activities at my backyard "patch" and on to the greater field of the bird's daily movements. It is in the complexity of the influences that work on each bird as it feeds that we find the answer to why they move around so much.

To the casual observer with a cup of coffee and a day's work to be avoided, the birds at a feeder are merely getting something to eat and fighting over it. But in fact, the birds are eating, avoiding predators, keeping warm, maintaining dominance in a social group of their species, establishing or losing dominance among the other species present, and doing any number of other things we don't understand too well. Some times of the year they're lining up mates and fighting off male competition. Other times they're gathering food for their young at a nearby nest.

In a neighborhood of patches, all of which tend to change by the week or by the day (if I go away for a week and the feeder is emptied after three days, for example), the bird with the best chances is the bird that conducts all those activities in a way that will allow it to keep as many options open as possible. Perhaps when a flock abruptly deserts a still-full feeder, they have filled up for the moment (they must get full sometimes, if only for a few minutes) and the best thing for them to do is to spend a little energy investigating the status of another patch that has been reliable or is getting promising. It may not be satiety that moves them on; it may be some other trigger.

Whatever the cause, the effect is that they keep moving and keep up to date on the status of many food sources. They don't do it only to satisfy hunger; a flock that moves many times a day isn't a flock that a predator will be able to key in on for a reliable daily meal. (Everybody is a predator in this situation; it may make the process easier to appreciate if you consider that the seeds in the feeder are no less prey than are the birds that eat them.) For that and other reasons, the movements of the birds give them the best odds of survival over the long haul. Not only are they harder to target that way, but they're more likely to come upon new food sources, new shelters, new nesting sites, new mates, or whatever else they need.

One of the small joys in a world where conservation biologists are almost hysterical about the disappearance of species and habitats is the kind of accidental diversity that we create around our houses. We labor with our picket fences and our lawn mowers to establish control over little rectangular territories, and here come all these animals—birds, squirrels, cats, dogs, insects—who superimpose their territories on ours.

There is something comfortingly sloppy about the way they ignore our boundaries, or treat them so differently. They put their own boundaries at all sorts of cockeyed angles to ours. They take something that we might prefer to think of as a sterile legal entity—the property line between our houses—and turn it into a delicious assortment of microhabitats. They take the conservative, stiff-necked grid of town lots and drop upon it a kaleidoscopic jigsaw of edges and centers, a mosaic composed of rhomboids, paisleys, and irregular nonagons that only they recognize, and only they know how to change.

In the early 1970s a University of New Mexico doctoral candidate named Stephen Beckerman, an anthropologist, spent three years studying the Bari tribe of the northern Andes. They are hunter-gatherers, and he accompanied them on many trips they called "hunts." To his dismay and confusion, these hunts rarely resulted in game caught, so rarely that he began to refer to them as "pseudo-hunts" and wonder why the tribe seemed to violate the conventions of foraging theory. In much less time, they could have gone to the nearest river and collected ample fish meat—why waste so much time trying to catch a few mammals? It seemed to him as silly as my fickle finches seemed to me when they flew away from a well-stocked feeder.

But eventually he noticed that a hunt was really a survey. As they went down the trail, the men were watching fruit trees to see what was ripe; they followed the stream to see where the fish were concentrating at that day's water levels, and checked the trails for droppings of tapir, monkey, and other mammals. On an even more fundamental level, they were maintaining important trails that would probably have grown over quickly in that thick forest. They kept moving, and by moving they improved their likelihood of long-term survival even if they didn't bring home a nice fat agouti each day.

It appears that any animal that has to make its living from nature has to have some pretty sensitive triggers to engage its movements, like whatever is the impulse that makes the birds move. For want of a better term, I'll call it a feeling. What does it equate with in humans? Some form of restlessness, I imagine. Does the bird instantly know when it's time to move on, or does a sense of dissatisfaction with things creep over it?

Is it anything like the human restlessness we call boredom? Boredom seems, on its most elemental level, to be a response to inadequate, disappointing, or insufficient stimuli. When I've been sitting in a business meeting for too long (ten minutes will do), and gradually I am consumed by an urge to go forage some beers somewhere, am I responding to some evolutionary descendant of the urge that moves the bird?

I like to think so. It makes my coffee breaks seem so much more respectable, knowing I can justify them in instinctual terms.

Songs of the Border

I lived for a couple of years in a small Pennsylvania town that was almost unrealistically picturesque, as neatly groomed and perfectly laid-out as those little plastic villages you buy for your electric trains. It was as if someone mowed every lawn every night, and assigned photogenic children to play cheerfully here and there after school, and (I remember this best) stationed all the most appealing birds to sing their finest songs just at dusk, when all the walkers were out enjoying the air.

There were streets in my little town on which, come summer, it seemed that every fourth or fifth housetop was occupied by a mockingbird in full song. As I walked I was always within hearing of at least two, and sometimes more, of these singular performers. I never tired of them. Even my limited knowledge of bird song permits me to recognize several species represented in the bird's repertoire.

Ornithologists have quantified the mockingbird's gifts. One mockingbird was observed to sing thirty-two identifiable imitations of other birds in ten minutes; another imitated fifty in an hour. Others have imitated a dog's bark, a wheelbarrow's squeak, a postman's whistle, the drumming of a woodpecker on a tree trunk, and even the tinkling notes of a nearby piano.

For the amateur naturalist this is charming stuff, in itself enough of a delight to make for a satisfactory evening of nature appreciation. Wherever I live, my evening walks are times of concentration on the day's doings rather than on what I see around me as I walk, so it wasn't until I saw the birds neatly regimented on their housetops, organized necessarily by the order of the streets, that it dawned on me that they were almost certainly singing for some good reason.

I'd never assumed, as so many schmaltzy songwriters seem to have, that these birds were singing for my edification; the songs were just so good that I hadn't thought to wonder what was really going on. But the uniform spacing of the birds, each separated from the rest by a few house-lengths, was just too obvious, and my curiosity finally kicked in. What we had here, I realized, was a classic set of territorial boundaries imposed on the human property lines. While other animals apply boundary markers with a variety of scrapes, scratches, and scent marks, some birds post their warnings with melodic no-trespassing signs that work just as well.

Wild animals have a much more fluid—and, I find, stimulating—sense of boundary than do humans. A songbird, except when it is migrating, may have a fairly clear area that it uses to satisfy its various needs for food, shelter, and mating. Radio tracking and other types of observation have shown that you can draw a reliable map of this area; people who draw such maps refer to the area as the bird's *range*. I was talking mostly about ranges in this book's essay, "Keep Moving." But a range is not the same as a territory, at least in the parlance of ornithologists. A range is simply an area used by the bird. A territory is a part of the range that means so much to the bird that it takes some form of possession. As the great bird behaviorist Margaret Nice put it more than fifty years ago, "The very essence of territory is its exclusiveness; if it is not defended it is not a territory."

Unlike human boundaries, which are as permanent as laws and real estate agents can make them, animal territories shift and fade; many animals are only territorial during mating, while others develop territories around a good food source and live their entire lives in a small, comfortable space that provides them all they need in food and shelter. Birds that are violently protective of their nesting area (whether it be the backyard that surrounds a songbird's tree or merely the portion of limb upon which a heron builds its nest in a crowded rookery) may later in the year group together for periods of close, placid companionship.

Territorial boundaries are also varied in their defender's discrimination. A male bird may defend its territory only against other males of the same species, while the male of another species takes on all comers, from humans to alley cats. A flock of birds may unite to drive off a common enemy.

The reasons—some proven and some only proposed—for all this elegant variation in territorial needs are almost as numerous as the variations themselves. Maintaining territories makes the process of pair bonding between mates that much simpler. The organization of available habitat into discrete "owned" units lets all the individuals involved know just where they stand, thus reducing the amount of effort put into fighting; an intruder knows where it can go unchallenged, and where a fight can be expected, so fewer fights occur. Birds spaced in the available habitat may be less vulnerable to rapid transmission of disease and genetic complications of inbreeding. Predators can't get more than one animal at a time when the prey is so well dispersed.

It is often true that we don't know exactly which of these factors are in effect, or if all of them are in effect at once, or how they work in concert, or if something else is actually causing the territorial displays of the birds we see. But there's no question that the robin beating his brains out on the hubcap of a car parked under "his" tree, or the hummingbird chasing his colleagues away from the sugar water at the window feeder, or the red-winged blackbird frantically buzzing a passing rowboat, is upset by some perceived violation of his preferred space. The only thing I will say for sure about these varied situations is that they're probably much more complicated than they first appear.

I recently realized that about my mockingbirds. Just as I was getting comfortable with the warm feeling that comes from knowing what's going on around you, I came upon a study of mockingbird territoriality that put me in my place.

Researchers in Florida, working with a highly visible and thus easily studied population of mockingbirds, discovered that the mockingbird was doing more than posting its land.

The idea behind the study is almost as interesting as its results. Researchers have trouble studying bird territorial behavior because birds are often hard to observe. Even if you can see a bird singing, you can't count on getting close enough to determine to whom (or at whom) it is singing. So even if you have a good map of an individual bird's territory, you can't tell for sure what its songs mean.

Scientists Randall Breitwisch and George Whitesides overcame that obstacle by studying mockingbirds on the relatively open terrain of the University of Miami campus at Coral Gables. The birds there were habituated to people, and it was relatively easy to observe them closely.

Breitwisch and Whitesides color-marked twenty-four birds and, among other things, kept track of the direction each bird's beak pointed each time it sang. What they found was a pattern of signal and countersignal even more involved than they had imagined.

Male mockingbirds without mates sang more than mated males, leading the researchers to suspect that there was more going on than boundary patrol. Unmated males apparently

sang not only to announce their territory but to attract females. (Some scientists contend that the mockingbird's vir-
tuosity in imitating other birds' songs is a device for impress-
ing potential mates, who are presumably attracted to the birds
with the largest repertoire.)

Both mated and unmated males did a surprising
amount of singing back into their own territories. For the
mated males, this may have been a way of keeping in touch
with the female on the nest. For the unmated males, broad-
casting their message in as many directions as possible, regardless of their location within their territory, let them reach
as wide an audience as possible.

Unmated males did not simply send out songs at random in all directions, though. If an unmated male's territory
bordered that of a mated male, the former might direct much
of his singing at the neighbor's nest, coveting his neighbor's
nestmate. There was good reason to think this was happening,
because mated males sing most while building the nest, and
then they quiet down. Perhaps the unmated male, recognizing a vacancy in the airwaves, filled it with his own tunes,
hoping to lure the female away from the nest when the time
came to raise the summer's second brood.

Mated males sang for the same reasons unmated ones
did; polygamy is not unknown among mockingbirds, and a
mated male, even though it sings less, may be keeping its
options open at the same time that it defends its present mate.

Curiously, males did not sing much when chasing in-
truders, although you'd expect them to be in full cry at such
a time, especially if the primary function of singing was to
maintain territorial integrity.

My housetop mockingbirds were up to much more than
I imagined. They sang to an amazing assortment of sounds
that incidentally came their way. They sang to each other,
and they sang to each other's mates. They sang because their

neighbor stopped singing, and they sang to absolute strangers. They sang to no one in particular, and to everyone who would listen. Judging from their reaction to my passing, they even sang to me, and they were so good at it that I still can't resist thinking that at least some of the time they sang just because it felt good.

Nervous Neighbors

It's surprising how little it takes to make grizzly bears nervous; you'd think an animal like that would be imperturbable. But then, nervousness has always worked really well for them. In a way, nervousness helped create them.

Steve French, research director of the Yellowstone Grizzly Foundation, and I recently spent parts of two late-fall days watching a grizzly bear guard a bison carcass. The bear was a four-and-a-half-year-old male, still young but more than big enough to impress any observer. I suppose he weighed 250 to 300 pounds.

Steve and I had a good observation position, slightly higher than the bear and a few hundred yards distant, where we were no disturbance to him. The carcass lay at the edge of a small aspen grove with a steeper hill just beyond. Two or three little draws led up the hill from the grove into a dense pine forest.

Steve had been monitoring this carcass for a few days— it was just off one of the park's unpaved secondary roads, already closed to tourist traffic because the fall rains made it too slippery for the average sedan driver—and had seen a variety of scavengers on it, including a black bear. The first day I went with him, I happened to have the carcass in view before he did, and I said something like, "Doesn't look like a black bear today, Steve."

Grizzly bears are as fond of bison meat as are most other scavengers, but by virtue of their size they can simply take over the carcass and chase off all other interested parties. They then begin to display what I call, for lack of a better term, nervousness.

Carcasses in Yellowstone are magnets for many appetites. Around any carcass more than a day or two old you assume you'll see evidence of ravens, magpies, and coyotes, and you look with particular concern for the tracks of bears (in case the tracks have only recently been emptied). Any time you're in bear country and you think you see a dead animal, don't just amble on up for a closer look; you don't know who's taking a nap nearby.

Often, bears will leave more dramatic evidence than the others. If the "long bones," like the biggest leg bones, have been shattered and worked on, a bear has been there; coyotes don't have strong enough jaws to do that. Bears also will bury the carcass.

Burial is an intriguing behavior. Some bears will dig at least a shallow hole, roll the carcass into it, then cover it over with earth. In the 1970s, there was a grizzly bear in Yellowstone known as Square Bear because he was observed digging a hole that actually seemed to have corners, like a grave. Other bears just pile dirt on top of the carcass until it seems like enough. In either case, portions of the carcass may be exposed, perhaps a leg or shoulder sticking up through the dirt—burial doesn't always equate precisely with visual concealment.

Burial seems to have more than one purpose, but I don't know which evolutionary priorities, so to speak, are behind the behavior. I suppose that sometimes it is most effective in smothering the scent and making the carcass a little less

easily locatable; a bear digs up so much earth that the whole
area around the carcass looks more like a bomb crater than
a place where someone is trying to camouflage something.
Such sites must be easy to see from a long distance, or from
the air.

Burial may also be a kind of warning sign, like scent
marking is for dogs—"This is mine, stay away." One of the
oddest examples of it I've seen occurred in a group of more
than a hundred dead elk, one of the few large groups of them
killed in the fires of 1988. There in the woods, with several
carcasses only a few yards away and dozens of others scattered here and there in all directions, a bear had taken the
trouble to partially bury two cow elk that had fallen next to
each other. There was no question that the burial was useless
for concealing either the sight or the scent of the elk, so perhaps it was a no-trespassing sign. Whatever the purpose of
such behavior, obviously it is deeply rooted.

It was well rooted in this young grizzly. By the time we
arrived, when he probably had been in possession of the
carcass for only a day or so, there was a shallow "moat"
(Steve's term) around the carcass, and piles of dirt over much
of it. He had dug no hole, so the higher he piled the dirt the
more obvious it was that something was there, covered up.

He was in the position he most often occupied, sprawled
on his belly right on top of the pile. Once or twice during the
several hours that we watched, he exhumed some portion of
the carcass and worried it, pulling off chunks of meat. But
most of the time he just protected it.

He had a very active definition of protection. A few ravens
were hanging around, sometimes sitting in the trees directly
above him, sometimes prancing only a few feet beyond the
moat, always looking for their chance. He eyed them with
anxiety, often turning abruptly to stare one off, or even taking
a step or two toward them. A coyote got the same treatment.

If the threat seemed persistent enough, the grizzly would

go back to digging, sometimes stepping toward the perceived competition, then back-digging with his forepaws rearward to the carcass. Steve told me that by a couple of days later, when I was no longer visiting the site with him, the bear had dug the moat so deep that he was now making longer excursions out between the aspen trees, and digging trenches back toward the carcass, pulling more and more dirt onto the pile.

All this proprietary activity reached its most exciting point—for us, at least—when the black bear showed up. The grizzly was resting on the pile, facing more or less northeast, when Steve suddenly noticed a black bear coming down one of the draws from the heavy woods behind the grizzly. The draw emptied into the grove maybe sixty yards behind the carcass, and the black was not yet in sight of the grizzly when we first saw it.

I must explain our excitement. Interactions between major carnivores are almost continuous on television nature shows, but in the wild you're lucky to see one once in a great while. Steve, with thousands of bear observations, had seen relatively few between these two species. I, with bear experience based primarily on years of reading obscure natural history and sporting literature, had read only anecdotal and often unreliable accounts of such interactions. For both of us, it was kind of a big deal.

I must also elaborate on the nervousness theme here. From the moment I had them both in sight, I was, with more analytical calm than I thought I possessed, also thinking about the long, convoluted evolutionary trail that had brought them here.

Bears and dogs have a common ancestor. They split in the mid-Miocene, about twenty million years ago. Before that, as the great European paleoecologist Bjorn (which means bear, by the way) Kurten (who has also written some outstanding novels about prehistoric times) has explained in *The Cave Bear Story* (1976), there are animals in the fossil record that could be described uneasily as bear-dogs, or, just as uneasily, dog-bears. After them emerged the dawn bear, *Ursavus elemensis,* a small, carnivorous mammal whose dentition (characterized by larger, flatter grinding surfaces better suited to vegetation) distinguished it as the start of one of nature's excursions into omnivory, an excursion that would result in the two animals now about to encounter one another in the aspen grove.

From the dawn bear's time, Kurten traces bear lineage through a series of paleontological *begats* to the mid-Villafranchian age, about two and a half million years ago, and the appearance of *Ursus etruscus,* the Etruscan bear, the common ancestor of several later species, including all three North American bears: the polar, the black, and the grizzly.

The Etruscan bear was a forest animal. The modern black bear *(Ursus americanus)* is much like it, with short, well-curved claws most useful in tree-climbing. Black bear sows protect their cubs by sending them up trees, and if the danger is grave enough they will follow them to safety. Black bears came to North America across the Bering land bridge about 250,000 to 300,000 years ago. They vary in color from all black to nearly blond.

The modern grizzly bear (*Ursus arctos*) became what it is by adjusting to new habitats. Dr. Stephen Herrero of the University of Calgary, who has done the most penetrating analysis of the development of modern bear behavior, describes the grizzly as "a forest bear remodelled to exploit the seasonal productivity of periglacial edges." The grizzly left the woods and became an animal adept at surviving in open ccuntry, from ocean beaches to alpine tundra. To do so, it developed long, straighter claws and massive shoulder muscles for digging. Its access to more kinds of food (it didn't abandon the forest entirely) allowed it to grow larger, and it became a more effective and aggressive predator. A younger species than the black bear, it was prevented by continental ice from reaching what is now the United States until 10,000 or so years ago. It too comes in many colors, and was named for the grizzled appearance of its coat, with various combinations of lighter and darker hairs and underfur giving it a more complex coloration than that of the black bear. Some early writers quickly mixed grizzly with grisly, complicating the name.

The polar bear *(Ursus maritimus),* by the way, is younger still, appearing in the fossil record not much more than 100,000 years ago. A fast-changing species that has reverted to carnivory from omnivory, it shows changes today even from fossils only 10,000 years old.

There are important behavioral consequences for the divergence of the black and grizzly species. The grizzly, without the shelter and security of trees, became a more aggressive defender of its young (who thus learned similar behavior). If you are a mother grizzly with nowhere to hide your cubs, or your food, from enemies, you tend to be a little nervous, and you tend to react a good deal more assertively than a mother black bear might when a threat appears. If you are the product of tens of thousands of years of mother grizzly bears doing this sort of thing, you're likely to be pretty good at it.

As Steve and I watched the black bear work its way down the draw, I got my first chance to compare these two species in real life, as opposed to in memories and photographs in which they were always apart. Telling the two species from each other is an important matter in bear country, and I had more than once written or explained that it was especially hard when only one was in view. I'm sure many Yellowstone visitors go home convinced they've seen a grizzly when they've seen a brown black bear, and far too often hunters in the forests near Yellowstone Park make the same mistake and shoot a grizzly.

There was no serious comparison this time. The black

was an adult of unknown sex, about the same weight as the grizzly—250 to 300 pounds. It was dark brown, and though it had all the general traits of bearishness—full deep belly, heavily muscled limbs, and the like—it was immediately recognizable as anything but a grizzly. The grizzly, with its more massive head, its pronounced shoulder hump, and its brindled coat of no describable single color, was nothing like the black. These were neighbors, not cousins.

The black bear obviously knew the carcass was there, and was on its way in to check it out. Steve said this wasn't the one he'd seen a few days before, which made us wonder just how many bears were involved in this picnic. When the black left the draw, it saw the grizzly and stopped to consider this dismaying development.

The grizzly, spread-eagled comfortably on its carcass pile, regularly lifted its snout to the wind, just to see what was new, but it took it a while to sense the black. In the meantime, the black, clearly disappointed at the presence of the other bear, began to feed—or pretend to do so—on vegetation along the hillside.

When the grizzly did realize the black was nearby, it reached new heights of agitation. In two days, we watched it react to the apparent threat of the black bear four times, and each time it did so a little differently. It might dig, it might pace, it might go to a nearby tree and stand against it, scratching its back on the bark. But each time, some threshold of turf and tension was finally crossed, and it would go for the black.

Don't ever try to outrun a bear. I know they look ponderous, even when they're under way, but they can lope faster than you can sprint. The grizzly, once he decided he'd had enough of this interloper poking around in the grove, would suddenly break into a gallop—though I'm sure it wasn't his top speed—and cover the distance almost as fast as the black bear was able to reach the nearest aspen and scramble noisily thirty or so feet up.

There was never any question of who was dominant. The black fled the instant the grizzly charged and, sitting on some heavy branch perch well up in the tree, watched the grizzly finish its charge and then wander back to the carcass. The outcome of each charge was so certain that even before the grizzly had reached the tree, he would merely be jogging along, as if tagging the trunk were just part of the ceremony. He would reach the tree, lift his front paws onto it, not fully standing upright, and turn away. By the time his paws touched it, his head had already turned. Only once did he even bother to look up at the black, to growl a warning or give it a nasty glare. Once he even turned his back to the tree and had a little back rub on the trunk, sure beyond any question that the black bear would not take advantage and jump him from above. Then he stalked off with an "I'm bad" sort of gait, to resume his pile riding.

The black would wait a few minutes, eyeing the grizzly through the intervening branches, then scoot quickly down the tree. Both days, its first descent soon brought a second charge from the keyed-up grizzly, followed by a second treeing, and then the black was allowed to leave.

It's not always this straightforward when the two species meet. There are a number of accounts, including some from Yellowstone, of black bears standing their ground against grizzlies. While we were watching all this, it occurred to us that it would be really interesting if a big male black bear were to come in.

That remark begs for a short concluding digression. In the informal jargon of Yellowstone bear study (and I suppose elsewhere), there are adult females and adult males, but there are also "big males." Once females reach adulthood, they tend to invest their energy in cub-rearing rather than in additional growth, so they more or less stop growing. But males have many reasons for wanting to get bigger and bigger, including their need to be able to win a female and hold her long enough to mate. The result is that the bear population winds up with a certain number of really large dominant males that may have a corner on the gene pool and certainly have a corner on anything else they want. When we see one of these we refer to it as a big male, though mentally I always picture the term as Big Male or, better yet, Big Male, Sir. Steve tells me that in all his observations of grizzly bears he has yet to see a big male run, either to or from something. He assumes they can; they just never need to. Maybe it's possible for even a grizzly bear to outgrow nervousness.

What we were wondering was how it would go if one of the black bear population's largest males were to come in now, and discover that the carcass was held by a young, inexperienced, and, within his own species at least, subdominant grizzly. I suspect we would have witnessed new heights of nervousness.

But it didn't happen, at least not while I was watching. My last sight of the bear that second day had him back on the pile, guarding his prize. A still photograph of him would have suggested a bear dozing off, but even a minute's steady watching belied that image. His eyes might close, and he might seem to relax, but regularly, every few minutes or so, his head would come up, those little eyes would scan the foreground, and the nose would test the wind.

Getting the Drift

The more I think of it, the more this sounds like the opening scene in a horror film. The heat of the August afternoon has finally subsided in the remote New England village. The sun has set and the moon will not rise for two hours yet, so the early night is unusually dark. Lights are winking out in homes all over town, and the filling station out on the state road has just shut off its big sign. The Ferguson place, a rambling, long-porched old house, sits dark at the end of Oak Street, the porch swing moving slowly in the warm breeze. But out back, past Judy Ferguson's rose garden, in the creek, something is just waking up. Something is living in Baldwin Creek.

Just under the surface, along the dark side of an undercut bank, a tiny, clawed arm appears from a rock crevice. Then another—and another. Driven by unknown urges, the creature looses itself into the swiftest part of the stream's current, unable to see that others have crept out to do the same. At first dozens, then hundreds, then thousands appear. Something has told them: It is time. Baldwin Creek, where Katie Ferguson washed her trike that afternoon, is thick with sharp-clawed forms, tumbling with the confused pulse of the stream flow.

But there the screenplay stumbles. All this really does

happen, but the assembled creatures don't emerge to devour the Ferguson's Irish setter (or the Fergusons). Most of them are less than half an inch long anyway. They simply drift, each one for fifty yards or so, then settle back to the stream-bed. The moon comes up, the movement subsides, and then it resumes again for a while just before dawn.

The creatures are the immature forms of a great variety of aquatic insects—mayflies, stoneflies, caddisflies, true flies, and others—and they have just participated in one of the least appreciated of insect activities. It is called behavioral drift, and it occurs in most streams supporting these insects. It wasn't even formally discovered until about twenty-five years ago.

There are many kinds of "drift" in a stream. Any time you put your nose to the surface of a stream clear enough to study, you are likely to see some insects rolling along in the current. Some have slipped loose from poorly chosen holds on rocks, others are working their way to the surface or the shore, to emerge as adults. Water forces its inhabitants downstream whenever they leave the protection of rocks and eddies, whether intentionally or accidentally. But behavioral drift is something more organized than that, and something infinitely more intriguing. It is the voluntary mass launching of huge numbers of these little animals into the current, which represents a displacement of a significant portion of the stream's population.

For many years, fishermen, at least a few of the more observant ones, knew that for some reason or combination of reasons fish in streams became more active, and did more feeding, just at dark. Some attributed it to the reduced light, which made the fish less cautious. One or two, though, wrote that aquatic insects, known generally as nymphs, seemed to be more active then too. In the early 1960s, scientists in Sweden, Japan, and the United States began to document what is called the "diel periodicity" of invertebrate drift patterns. Most important, they reported that the drift was much heavier at night and was clearly influenced by light intensities. Up to then it was only assumed that there were always some insects loose in the current; now it became apparent that the movements had a pattern.

Eventually, many patterns were found. On long winter nights, there might be as many as three peaks of drift, alternating with periods of little movement. In summer, when nights are shortest, only one such flurry of activity might occur. The vast majority of the insects involved were those closest to maturity, what might be called the grown-up immature insects. The whole process was always triggered by a lowering of light levels; a bright moon could retard the drift, causing the animals to settle to shelter, and similar effects were observed when artificial light was shone on a stream.

An ecological phenomenon with only a brief scientific history is usually not a fully understood phenomenon, but there is consensus that behavioral drift is the result of population densities. When an area, whether it be big-game range or prairie dog habitat or a streambed, reaches the limits of its ability to support the resident life forms, something has to give. Nature's reactions to a range that has exceeded its carrying capacity are many and include widespread starvation, but streams provide a handy alternative to dense population—just let go and drift downstream to greener pastures. This, according to most modern theorists, explains behavioral drift. Under cover of darkness—presumably the safest time for the hordes to expose themselves to their predators—the immature mayflies, caddisflies, and the rest reduce population density by leaving.

We miss so much. Most people don't even know that their streams are inhabited by this richness of life in the first place, much less that thousands (and, in larger streams, millions) of these delicate little animals spend their nights leapfrogging downstream in search of more room.

Imagine that you lived in a world with an atmosphere

so dense and forceful that you could be carried by it. Our atmosphere has to reach its most catastrophically energetic state to lift us. We can easily stand straight against a thirty-mile-per-hour wind, but try to stand knee-deep in a stream that is moving even eight miles per hour. Imagine that your ancestors evolved in the face of that kind of pressure. If you were a fish you would be streamlined, perhaps tubular, perhaps flattened, and you would spend most of your life swimming just fast enough to stay in the same place. If you were an insect, you might have evolved any of the thousands of forms insects now have.

But the current would be more than an adversary, just as the wind is for us. You would not fight it so much as cooperate with it; never could the phrase "go with the flow" have as much meaning for us as it does for the life forms of such a currented world. The flow would bring you food, it would dig you shelter, and, when the neighborhood got too crowded, it would carry you away.

The Odd Couple

In many years of happy, insatiable, and blithely undis-
ciplined reading about western wildlife I occasionally
came across casual, almost peripheral, references to an odd
and not well understood relationship between two predator
species, the coyote and the badger. Every couple of years,
while I was looking up something else (most good stuff is
found while looking up something else), there would come into
view another incomplete, vagrant record of a coyote and a
badger "hunting together."

Mind you, I was not especially impressed. I, after all, was
more enlightened than the naturalists of the 1930s and
1940s, who were infamous for their anthropomorphism. These
were the same people who made plant-eating animals speak
with the voices of precocious children, and gave meat-eating
animals the personalities of barbarian warlords. It was instantly obvious to me that they weren't seeing a coyote and
a badger hunting together. They were seeing a badger hunting,
and a coyote hanging around to catch all the mice and
squirrels that the badger couldn't.

It was classic coyote opportunism, the great Trickster
at his best, keeping just out of reach of the badger and

95

watching it dig out squirrels and mice. As soon as the little animals got worried enough by the badger's excavations that they raced out of one of their burrow's back doors, the coyote was there to nail them. This was no partnership; this was theft. The badger, not being as fast as the coyote, certainly hadn't volunteered to participate in this arrangement; it just couldn't do much to stop it.

But a few years ago I was giving some friends an auto tour of Yellowstone Park, where I have lived now and then over the past two decades. We were driving down a busy park road to the drone of my informed narrative when up ahead, on a hillside just off the highway, we saw two gray shapes bustling along. They were in view for perhaps twenty seconds, close enough for us to see the jaunty lope of the coyote and the trundling gait of the badger, who was right behind him.

I was so excited at our good fortune in seeing this rare combination of creatures ("This isn't just some more elk, you know—this is important!") that I didn't really think about it until a little while later, when it dawned on me what I had just seen. The badger was *right behind him.* The badger wasn't being shadowed by the coyote, he was voluntarily following the coyote.

So much for my enlightened perspective. The revelation sent me back to the coyote literature, where I discovered what I had missed before: that badgers have indeed been seen following coyotes, just as coyotes have been seen following badgers. Since then I have spent an embarrassing number of hours trying to sort this out. It has become one of my favorite things to puzzle over as I drive along. Badgers and coyotes really do seem to get something out of hanging around together, and they do it with mutual agreement.

But what is it? The coyote's rewards are pretty obvious, but nobody has ever seen a coyote catch a mouse and then go share it with the badger.

The people who had thought and wrote about this relationship before me had a lot to say about the possibilities. The great western folklorist J. Frank Dobie, probably under the healthy influence of various native American mythologies (in which the coyote has considerably more stature than our culture is inclined to give it), decided that we are not going to figure this one out, that the relationship between the coyote and the badger is "not sensible to civilized man." Dobie said that we can "err by trying to assign utilitarian motives to all behavior of other animals." The two may keep company, and play together (as they have been observed to do), just for the sake of keeping company and playing together.

I find that satisfying, if not altogether sufficient. Coyotes are sometimes found in packs, and maybe a solo individual will go interracial in search of companionship. But badgers are not known for making friends except with other badgers, and then only long enough to make new badgers. It is true, as Dobie proposed, that animals of different species will occasionally pal around together for no better visible reason than that they want some company. The attachment can be a deep one, as shown by the companionship bonds formed by various domestic animals. But that possibility, that the coyote and the badger could just like each other's company, does not really explain why it happens in the wild again and again. It doesn't seem that such a friendship would occur in the wild as often as it does without some other factor contributing, something that would get them together in the first place and keep them together long enough for the bond to form.

So I still suspect that the initial reason is more fundamentally survival-oriented. Maybe the coyote really is just taking advantage of the badger, but after doing so for a few weeks the two of them get used to it and a bond forms (sort of like the involuntary bond that forms between some kidnappers and their hostages, I suppose).

Or, maybe the coyote does bring something to the relationship. Dobie thought that the coyote might contribute by chasing squirrels into their burrows so that the badger could then dig them out. The coyote might have some advantage over the badger, in terms of being able to see farther because it's taller, or simply in terms of better eyesight, which enables it to bring to ground more squirrels than the badger could alone. Then both animals might benefit through the random distribution of the cornered animals, some of whom flee from the digging badger into the coyote's reach, others of whom are dug out by the badger. Perhaps some of the cornered animals don't flee from the badger because they see or smell the coyote waiting just outside the back door. That would certainly benefit the badger.

In the last year or so, I've turned this casual, recreational-thinking topic into something a little more serious. I've been collecting all the accounts I can find, historical and modern, of the two animals together.

Some modern ecologists have offered me intriguing thoughts and observations on all this. Dr. Robert Crabtree, who has done some of the most important and revealing coyote research in recent years and is currently studying Yellowstone coyotes, suggested that one of the ways the bond might form is during the youth of the animals. Coyotes and badgers den in similar terrain, and a pup and a kit might encounter each other while still playful and just looking for some company. Bob doesn't suggest that it happens that way a lot, but it makes sense that it could easily happen that way. In fact, it appears to me that it probably happens in a variety of ways, with a variety of results.

Consider a sighting reported to me by Dr. Ernest Ables, a wildlife biologist at the University of Idaho. He and a group of students, on a field course in Yellowstone, were watching some elk one day when a coyote happened by. As they watched, they realized that the coyote was being approached by a badger. Dr. Ables' own description does the event justice:

> When the badger was about 20 feet distance the coyote went down on its front feet with its hind quarters in the air with tail wagging. This is a greeting behavior many of us have seen in wolf movies and which is common in domestic dogs, especially subadult dogs. As the badger approached close enough for physical contact, the coyote rolled over on its back, kicked its feet up and down then ran around the badger in what can only be described as a playful romp. During this time the badger moved its head up and down a few times but otherwise showed none of the greeting behaviors of the coyote. Both animals then proceeded in the direction that the coyote was initially travelling with the badger sometimes in the lead and the coyote sometimes in the lead.

Dr. Ables, after watching the two hunt ground squirrels, concluded that "these two animals knew each other and had hunted together before."

On the other hand, some pairs don't seem nearly so at ease with one another. One of the special treats of searching out badger-coyote stories was getting to meet Bob Landis, a skilled Montana nature-film maker among whose outstanding television films is *Song Dog*, a remarkable portrait of coyotes. He graciously invited me to view some of his unpublished footage of coyotes and badgers, and it was the high point of my search, perhaps even more exciting than my own sighting. Bob has accumulated extensive footage of the two together, wandering across sagebrush flats, separating and joining and separating, and, always, looking for food.

What struck me most about the pair Bob filmed was how

different their relationship was from that described by Dr. Ables. Bob's pair had a much more uneasy détente. The coyote was not frisky around the badger; in fact, the coyote was nearly obsequious. At one point the coyote inadvertantly approached the badger too closely from behind, and when the badger suddenly realized where the coyote was, there was a swift turn and a sharp snap of teeth at the flustered coyote, who seemed to have meant no harm. Even when the two bedded down, the coyote surrendered the shady spot it obviously preferred to the more assertive badger, and only approached and lay down with greatest caution.

Though these and several other accounts I've been given make it clear that the two do hunt together, all the other questions remain unanswered, or at least partly so. Conjecture leads us to assume that both animals get something from the company, but I suppose it's possible that the badger just gets used to the coyote and never really expects much from the deal.

The extent to which hunts are truly cooperative—that is, mutually understood to be cooperative—is also difficult to measure. Biologist Kerry Murphy, who is currently studying mountain lions in Yellowstone, was traveling with some friends along another stretch of the road where I saw my pair, when they too saw a badger following a coyote. This time a second badger appeared and tried to approach the one with the coyote, but was rebuffed. It also seemed to have been injured not long before.

The coyote and the remaining badger continued on and engaged in what Kerry thought was "a possible cooperative hunting effort":

> The coyote, leading the badger, stopped in a semi-crouched and alert position near a 0.75 m. × 0.75 m. bush and waited for the badger to approach on the opposite side. It appeared that the coyote was

waiting for the prey to emerge from the bush. The pair was last observed travelling westward together.

Badgers have a reputation for unsociability that is surpassed by that of few other wild animals. I imagine that it is overstated, but it is based in part on long human experience with hostile, ill-tempered badgers. At first glance, going by the prevailing stereotypes, the industrious, cranky badger would seem an unlikely candidate for friendship with the crafty, larcenous coyote. Of course, those are stereotypes, not real animals. The truth is that the two animals are thrown together by their interest in the same prey on the same ranges.

What goes on when they meet and team up really is, as I told my friends that day a few years ago, important. In fact, it is an extraordinary event in the natural world. Hardly anywhere do two predators cooperate like this. Even if we cannot yet determine the extent to which the relationship is truly cooperative, and even if we don't understand how it happens, we can't deny that it happens again and again. Apparently the two animals encounter one another often enough, and (either by accident or design) find worth in sticking together often enough, to make the arrangement work.

There is no reason to believe that each pair has gotten together for the same reasons, or that each pair has exactly the same mutually beneficial relationship. It already looks to me, based on the accounts I've found, like there may be considerable variation from pair to pair, not only in how they treat each other but in how they hunt.

Ultimately I may have to admit that Dobie was right, and this unusual animal behavior is not "sensible to civilized man." But I'm a long way from letting go of this one. It's just too engaging, and too much of a challenge. In the badger-coyote relationship we have a real anomaly, in which two competing predators with substantially different behavior and physical equipment team up, rather than duke it out, to get food. Some peculiar combination of behavioral traits and environmental conditions has led them to this, and I suspect that if I can just gather enough variations on the story, that combination will make more sense than it does now. It must—it obviously makes a great deal of sense to the animals.

All Together Now

The slickest choreography I've ever seen has been performed by minnows. I've watched schools of them cruise the shallows of lakes, and the quickness with which they could change direction, speed, or depth in seeming unison was so mechanically precise that it was as if they had one mind.

Then, a few years ago, I happened to see a flock of gulls corner a shoal of small fish—I don't know what kind—in shallow water along the northern California coast. The birds had the fish nearly to the beach, in only a few inches of water, and were in full frenzy. Small fish appeared in shiny sprays as a gull burrowed along just under the surface, and the water churned as other gulls slapped at the fish from above. That set me to wondering why all those fish would want to hang around together in the first place and set themselves up for such mayhem.

It used to be thought that schooling, herding, and flocking were functions of some simple social drive; animals formed groups because they liked the company. The social animals, that is—others were solitary. Of course, that simple an interpretation would not do. It was enough for some to notice that the social animals and the solitary animals could most of the time just as easily be called prey and predators. From then to now has been a halting, often argumentative progress to the realization that these groupings have many functions beyond providing like-minded animals with good company.

Animals join up to protect themselves from predators, of course, but protection can mean many things. It may mean that the animals, as a group, can defend themselves better. It may mean that each individual is there only because the larger the group, the smaller the likelihood that he will be the one eaten. It may mean that by grouping up, the prey force predators to travel greater distances between kills.

On the other hand, some animals group up *in order to* prey. Small schools of predators—consider killer whales, wolves, and striped bass—may be more efficient as a group, both in finding prey and in taking it. Everybody seems to have evolved in response to everybody else.

There are other reasons for the formation of animal groups, though. Fish that travel long distances during their lives may need to stay together so that at mating time they can find partners. Younger members of a bird flock may benefit from the group's accumulated experience along an established migration route.

Many animals may even find travel physically easier in a group. Schooling fish can "ride the edges of vortices made by other school members in front of them," as one biologist put it, thus working less. That sort of "slipstream" trick is also practiced by many birds, who follow off to the side and immediately behind their fellow flock members.

What is most intriguing to many of us who like to watch all of this, though, is that choreography I spoke of earlier. How do umpteen hundred small birds crowd into a small space of sky and avoid turning into a demolition derby every time someone up front changes his mind?

Naturalists have been working on this one for a long time. In his classic book about life on Cape Cod, *The Outermost House* (1928), Henry Beston asked the questions about as well as they've been asked:

> By what means, by what methods of communication does this will so suffuse the living constellation that its dozen or more tiny brains know it and obey it in such an instancy of time? Are we to believe that these birds, all of them, are *machina,* as Descartes long ago insisted, mere mechanisms of flesh and bone so exquisitely alike that each cogwheel brain, encountering the same environmental forces, synchronously lets slip the same mechanic ratchet? Or is there some psychic relation between these creatures? Does some current flow through them and between them as they fly?

The possibility of psychic relations and mysterious currents has charmed many a birder. Everything from electromagnetic impulses to bird telepathy has been proposed to explain this amazing close-formation flying.

Some intriguing work has recently been done by a researcher named Wayne Potts at the University of Washington. Potts used high-speed photography to capture the movements of flocks of dunlin (a sandpiper-type bird) on Puget Sound. He recorded what he called a "manoeuvre wave," which was initiated by one or a few birds as they changed direction. All that was required for this event was that a bird somewhere on the edge of the flock turn *toward* the flock; if the individual moved away from the flock, the other birds would not react this way.

Potts called it the "chorus-line hypothesis," and here is how it worked: As the bird in question was caused (by a predator or some artificial alarm created by Potts) to turn in toward the group, the birds nearest him also turned, if only to avoid collision. As the "wave" spread from that one bird, the so-called chorus line took effect. Those first birds could only move as fast as their reaction time would allow, but the birds throughout the flock, even those far from the source of the wave, could see it coming and coordinate themselves to change direction more quickly. Thus the wave accelerated across the flock, permitting most of the birds to respond to the change in direction far more quickly than if they had been reacting only to the behavior of the birds immediately adjacent to them.

It's called the chorus line because other studies have shown that a human chorus line, taken by surprise by a maneuver initiated at one end of the line (imagine a wave of high kicks, for example), could move the maneuver down the line twice as fast as the dancers' known reaction times would seem to have allowed. They could do it so fast because they did not wait until they saw the person next to them start to kick, but anticipated the wave of kicks when it was still several people away.

Researchers have observed that schools of fish act similarly to flocks of birds in that there is no "consistent leadership" in a moving school. Those in the front are suddenly on the side, or in the rear, when the group changes direction.

This makes a lot of sense, because it means that the group does not have to rely on a single leader to get it through a direction change. If it did, a school of fish would turn by going in a big circle, sort of like an aircraft carrier ponderously changing directions. Worse, if a predator drove into the flock from the rear, fish waiting for a signal movement from a lead fish might be dead before the lead fish even knew there was danger. Instead, the school can instantly—almost automatically—go in a new direction, away from the imagined threat.

In practice, this sometimes means that the school simply disintegrates, as all the fish go into a "big bang" exercise, each one headed away from the attacker, with little evident regard for the others beyond trying not to run into them. If there are several attackers approaching from several directions, there may be a series of little bangs, after which, based on my casual observations, things get so chaotic it's impossible to see any group coordination.

One of the other big puzzles involving this coordinated flight is how it starts so instantly. When I suddenly open my back door, how do dozens of sparrows in the apple tree all manage to take off at once, when they can't see each other and are making such a continuous racket that no alarm signal would seem audible over it? And who decides which direction they will go (they always seem to leave knowing where they're going)?

A University of Rhode Island zoologist named Frank Heppner has been working on computer models of just this situation. He spent years trying to develop a reasonably tidy computer model to explain flock movements, before some colleagues persuaded him that the recent scientific interest in chaos might help him out.

I won't pretend to have a good grip on chaos theory. Even *Science* magazine recently admitted that it is "a mathematical concept that is rather difficult to define precisely." Its point in this case, however, is that the traditionally accepted "laws of nature" are not up to fully explaining the incredibly complex happenings out there in the natural world. My own readings of the debate over whether or not chaos theory is a true scientific revolution suggest to me that the jury is still out, and that ecology is one of the fields where it has been least successfully applied. But it certainly helped Heppner.

What it seems to come down to for those interested in ecological processes is this: From the initial event, say the alarm that causes all of the birds to jump from the wire at the same instant, it is progressively more difficult to predict what those birds will do the more time passes. That may seem intuitive, but what chaos scientists are telling us is that the level of difficulty is almost unbelievably higher than you might ordinarily think. There are simply too many variables, and each of them is itself subject to too many variables of its own, to consider the results of the initial jump predictable with present scientific skills.

So it was that Mr. Heppner created a computer model— a simulation of the behavior of the birds as they left the wire. In order to try to replicate their behavior, he programmed four rules for them: 1) The birds are attracted to a "focal point," like the wire or a tree, and are more attracted to it as they approach it; 2) They are likewise attracted to one another, as long as they don't risk collision; 3) They want to keep moving along at more or less the same speed; and 4) They are routinely vulnerable to external influences, such as predators or wind, on their course of flight. Heppner recognized that there were probably many more rules actually in effect in the real world, but these made a nice start for this experiment.

Programming these rules into his model, he "released" a small flock of birds (in his diagrams they are little arrowhead shapes) from a wire and let the four factors take effect. Sure enough, the birds moved pretty much in a straight line for a moment, then broke that line into a loose mass, circled around, and came to rest in a nearby "tree." Varying the relative power of the four rules affects the way the birds move, so that they are anything but little mindless machines carrying out a simple act.

Neither Heppner nor his colleagues think he's duplicated bird behavior precisely, but he has shown that the movement of the birds is extremely complicated, and hardly predictable by the casual observer. In a recent interview in *Science News* he commented, "We're not saying that this is the way birds do it. All we're saying is this is the way birds *could* do it."

Which, I'm glad to say, leaves plenty of room for speculation and doesn't entirely disillusion those of us who are still waiting for ducks to reveal their telepathic powers. How, for instance, does a school of fish, if it's truly leaderless, decide where to forage next? They are susceptible to the same sorts of habitat and range limitations I've already mentioned. I suspect their needs are straightforward enough that there isn't a great deal of individual initiative or variation among them; they go where the currents, foods, and other general habitat conditions compel them to go, and they probably become accustomed to some range or set of patches that they revisit regularly. They certainly couldn't afford to have too strong a leader, or they would be in a real fix if they lost him.

You may have noticed that automobile traffic seems to move in similarly mysterious waves, slowing and then speeding up. Though most commuters would agree that the result is clearly chaos, something akin to the chorus-line maneuver seems to be occurring. Most of us "drive ahead," as the driver's ed. instructors used to put it. We watch—through windshields, up hills, around bends—several cars ahead of us. When a car up there slows down, we react immediately, without waiting until the car in front of us causes us to.

It appears that as traffic moves, say, along a busy expressway, these waves are constantly being built into the traffic flow as people slow to exit or accelerate from a ramp into traffic, causing other cars to shift lanes, speed up, or slow down. The difference with cars may be that they are not in distinct flocks but are a continuous, unevenly distributed body of objects. Because of that I suspect that a single action by a car way ahead of me may initiate ripples in the traffic flow that last long after that car has exited.

Cars, however, give us both advantages and disadvantages over groups of birds. They allow us to move faster than we could under our own power, but they also allow us to exceed our reflexes' ability to react correctly when the chorus-line wave comes rolling toward us in the form of some hastily decelerating Buicks.

At this point in my ruminations I find myself back to fish. Those Buick drivers remind me of the trout I saw in a hatchery in the Adirondacks not long ago. I was being given a tour by the resident biologist, who was caring for a batch of large indoor tanks, each holding a different strain of trout or whitefish. He pointed out a shallow tank containing hundreds of tiny trout "fry" (fish only an inch or so long). This particular tank was full of the offspring of wild fish, ones that had been captured from a stream and stripped of their eggs and milt so that the young could be raised in captivity and released back into the stream as young adults.

The biologist slowly moved his closed fist out over the tank, then suddenly flashed open his fingers, causing instant panic (or chaos) among the little trout, who scattered in all directions. These fish were instinctively spooky.

Then he moved to another tank, this one full of similar-size fry that had been produced from hatchery fish. They were the product of dozens of generations of captive breeding. Again he moved his fist out, and again he opened it abruptly, but the fish took no notice at all. It was a graphic lesson in how quickly reflexive behavior can be dulled or lost. Easy living and protection had cost them dearly in self-preservation skills; they had become as inept at the natural choreography of survival as the human freeway flocks, and once they were released into the real world, their losses would be even heavier than those of the incurable idiots who tailgate me with no concern at all for the elegant choreography of motions and waves that surrounds us.

Biological Storms

Healthy, natural landscapes often seem to produce life with profligate abandon. The most popular textbook examples in recent North American history include the fifty million bison that lived on the western plains in the early 1800s, and the estimated nine *billion* passenger pigeons that darkened skies for hours as late as the 1850s. These huge numbers of animals swept across their habitats with a force that did not affect the landscape so much as it *created* it. Vast prairie rangelands coevolved with the herds of roaming herbivores, flowering, taking seed, and growing in time with the annual "predation" of the seasonal migrations of the mammals. Similar effects were no doubt created by the seasonal passings of the pigeons, which the great ecologist Aldo Leopold once described as "a living wind." In fact, it was also probably Leopold who coined the term I find most satisfying as an evocation of the power of such faunal richness: "The pigeon was no mere bird," Leopold wrote, "he was a biological storm."

Alas, the storm has passed. A fenced and roaded prairie has no room for bison. People in the densely settled eastern forest would not welcome the thought of mile after mile

111

of pigeon roosts. We certainly didn't have to destroy these animals with such mindless delight, but they were probably doomed no matter how politely we might have chosen to go about settling their habitat. By 1900, the wild bison were all but wiped out (the herds would be rebuilt as needed for commercial purposes). The last passenger pigeon died in the Cincinnati Zoo in 1914 (her name was Martha). It is ironic but true that a species of animal that supports itself through vast simultaneous movements of huge numbers of individuals may be more vulnerable to annihilation than a species whose individuals are far less numerous but distinct and separate in their movements.

But there are still other storms, and in their more local way they are just as awesome. In their numbers they may be more so. Insects make them.

I first learned about them when I became a fly fisherman. Fly-fishing demands of its practitioners at least a modest knowledge of stream entomology because most of the "flies" you fish with are made to imitate the forms of emerging aquatic insects.

The emergences, called "hatches" by most fishermen, are often spectacular events. In a period of an hour or less, millions of the larvae of some insect form (most common are mayflies, stoneflies, caddisflies, and true flies) leave their safe shelter on the bottom of the stream and rise to the surface, where they shed their nymphal skins, fly away, and in a day or so mate and die. This sudden concentration of meat in open water attracts trout and other fish, who feed heavily and thus become vulnerable to the well-tied and well-cast imitation. The heavy "hatch" is one of nature's grandest pageants. It seems a shame that only a few entomologists and fishermen know about it.

The heaviest such emergence I've personally encountered occurred one evening along the shore of a giant lake in Nicaragua. Looking back, I'd guess the insects were true flies, billions of them coming off the surface of the lake and masking the headlights of the car as we drove along. We had to stop every few minutes and wash and scrape the windshield.

But heavy emergences are not limited to subtropically lush settings. From the Arctic to the high desert, anyplace there's suitable insect habitat, the swarms can come. The Mormons of old endured their "plague of locusts," and most parts of the country experience occasional population irruptions of this or that insect. This is true also for the aquatic insects I've been talking about here. On any summer night, along any reasonably clean lake or stream, there might be these sudden fogs of life swirling along the shoreline.

B. D. Burks' small classic *The Mayflies, or Ephemeroptera, of Illinois* (1953) contains a photograph first printed in the *Chicago Daily News* in July of 1946, of a man trying to clean thick gobs of mayflies, just emerged from nearby waters, from the bumper and radiator of his truck. At his feet, the bodies of the inch-long flies are piled half a foot deep on the road. The newspaper said they were clogging radiators and making the roadway slippery. An early California traveler, William Brewer, reported in 1863 that the brine flies in Mono Lake "drift up in heaps along the shore—*hundreds of bushels could be collected....*"

Such events are not uncommon, either in the historical literature or in today's world. A couple of years ago I was fishing the famous "Mother's Day hatch" of caddisflies on Montana's Yellowstone River. The half-inch-long, tan flies were so thick on the surface of the water that, as my companion put it, it looked as if someone had backed a big grain truck up and dumped it in the river. Saying there were millions of them would be an understatement.

This kind of fertility in a natural setting inspires different responses in different people. There are those who see it as a waste of life, as many of the insects will never mate or lay their eggs. There are those who see it as proof that nature often will produce a "surplus" for the benefit of other life forms (especially us, our benefit being such a high priority in so many people's minds). And there are those—including me—who are disinclined to load the phenomenon down with value judgments, but who find it all intriguing enough to try to imagine how such a thing developed in the first place.

Those Yellowstone River caddisflies, for example, were offering my pal and me a lesson in population survival strategies that we were reluctant to learn. Trout were rising by the dozens, eating the flies as fast as they could, but it was almost impossible for us to catch a fish. Anywhere we tossed our artificial flies, any square foot of water we hit with a cast, was already occupied by scores—sometimes hundreds—of living insects. The odds of a fish taking ours out of that whole crowd were tiny.

And, if the caddisfly population had had a voice, it would no doubt have told us that this was just how it wanted things. It might have wisecracked that we were such inefficient predators that we didn't matter at all. But it would also have pointed out that there were other predators, especially those trout, and the swallows wheeling over the river, who took a fair toll of the population, and that this fabulous "overproduction" ensured the survival of ample flies to mate and carry on the species. That there were many millions of survivors, rather than whatever absolute minimum might have been necessary to perpetuate the species, was no concern of nature's. Precision has never been nature's strong suit when it comes to population survival.

As I've already suggested, "safety in numbers" can be read more than one way. A herd of grazing animals, or a flock of small birds, may by its numbers be more intimidating to a predator, and thus less likely to be attacked. In that situation, numbers provide safety for the individual—better off with others of its kind.

But a mayfly does not protect itself from a trout by

hanging around with other mayflies. Trout love bunches of mayflies and like nothing better than gobbling down a whole clump of them at once. Being part of a mayfly flurry won't frighten off predators, but it still may help the individual mayfly survive. For the individual mayfly in that situation, the safety to be found in numbers is one of lessening the odds; the trout can eat only so many bugs, and the more bugs there are, the less chance any one of them will be eaten.

Here we reach the level of ecological operation most often of interest to population biologists. Besides the safety of the individual, there is the safety of the whole group. That is, if you want to make sure that the population goes on—that the species survives—and if you know you're going to face predation, one course of action is simply to give the predators all they can eat. Just make sure they can't eat all you can offer, and you'll survive.

Of course, the mayfly population has no group consciousness that "knows" what it must do to survive hungry trout. The storm of emergence is a response to environmental conditions, a response that evolved over time just as the trout and their appetites similarly evolved.

You, like me, may have been brought up on simplistic notions about the forest being a well-tuned machine. Nature, we were assured, is like a wheel: if you take out one spoke the whole thing will collapse. Under this view, nature (or Nature) has a grand plan and all these balances between predators and prey, between grazers and grazees, are as immutable as baseball ground rules; it's only when humans begin tinkering with them—killing off this predator or fencing that range—that the machine gets out of kilter and changes. Though professional ecologists long ago left that mechanistic outlook behind, it's still popular with the public.

Truth is, nature doesn't make value judgments about tidiness and order. Even without the meddling of humans, nature is always beating the devil out of the spokes, sometimes knocking them clear off the wheel. They don't call it wild for nothing.

The biological storms are a great example of just what a gloriously sloppy mess nature can make of a situation given the chance. Chaos, however fashionable it may be among the scientific conceptualizers, is old news to mayflies and bison.

We humans like to think of nature's balance as something static and predictable, something without violent changes in condition, but what we have in the biological storms is something far more volatile. Massive simultaneous movements of life forms, sudden "birth storms" in which a large herd of deer or elk all give birth to young in a period of a few days, a single wildflower releasing thousands of seeds, only one of which may take root ... these are not events somehow "designed" to maintain the status quo in a natural setting. They are, instead, events that can produce wild fluctuations in that setting, fluctuations that would test most people's traditional notion of what constitutes a balance.

For those with even a mild sense of adventure about what goes on in nature, there's a thrill in this, especially if you go to the trouble of putting yourself in the middle of one of these storms. I spent an evening on New York's Beaverkill a few years ago, fishing with a few friends. For an hour or so we stood knee-deep in the still river, fidgeting with the absurd assortment of tackle and gadgetry fly fishermen haul around with them. Every now and then a large white mayfly would appear, settle to the water, and disappear in a slurpy splash as a trout fed.

Then, just at dark, the flies came on in numbers, first hundreds, then thousands, then millions, thick in the air around us (fishing writers who describe these events as "blizzardlike" are not exaggerating), coating the surface of the water, driving the trout into a feeding frenzy. It was like throwing dead chickens to sharks. The inch-long mayflies, fluttering in our ears, crawling down our collars (they can't bite, thank heaven), and landing on our noses, were unsettling, and the wildly slashing trout were enough to unnerve a saint. But the raw, almost obscene flushness of the whole proceeding was what really made it memorable. For me, there was no question at all of some higher-law "balance" being in effect; when the storm was on, everyone was out to get the best they could, and everything was up for grabs.

When the hatch was over, it seemed as if the evolutionary goal of the fish was to eat every insect, and the direction of the insects was to smother the earth in deep drifts of mating bugs. You can call the result of all that energy and commotion a "balance" if the term pleases you, but I prefer to think of it more simply: as a period of calm between the storms.

Apologies to the Salmon

I am a salmon watcher. I stand in a coastal river in Washington, gaping at a mute, ragged, yard-long fish already rotting to death as it sways in the current in front of me. I reach out and touch it with my fishing rod, and it only drifts a foot or so farther from me in response, so unconcerned is it with my existence. I am no threat to it now. The patchy fungus on its sides seems to spread even as I stand there. Its journey over, spawning completed, the fish is dying.

Like so many observers before me, I'm inexpressibly grateful for the chance to be here and watch the conclusion of its astonishing journey. Here is a fish that has traveled from the gravel of a small stream down through the big river to the ocean (for some in the upper Columbia River, it's hundreds of miles just to reach the ocean), and then off to some feeding ground hundreds or even thousands of miles away, returning a year or several years later to spawn in the very bed from which it hatched. Only the dullest observer could miss the romance in that odyssey. I feel like I should say, "Welcome home."

For thousands of years, American Indians welcomed the salmon back to their rivers. For centuries, naturalists and fishing writers have celebrated the return of the salmon. Those who have written about the five species of Pacific salmon that come into the rivers on North America's west coast, where I've done my watching, could add a profoundly bittersweet touch: the death of every single one of those millions of marathon swimmers after spawning.

Nature appreciation, besides its emotional satisfactions, is an exercise in perspective, in learning how to stand back far enough to take a larger view. Stand back from the gravel bar and see the whole watershed, with fish moving to their respective home pools here and there in hundreds of miles of streams. Stand back farther yet and see the whole saltwater range of the fish, up the coast to Alaskan waters and back. It is this perspective that has most inspired writers with admiration for the fish's great migration and wonder at how it finds its way home.

But stand back even farther, far enough to see the whole Pacific basin, and an even grander perspective emerges.

Salmon are *diadromous,* meaning they, like numerous other fish species, move between fresh and salt water, a process that requires a great physiological change in the fish's body. So many fish do this that naturalists have long wondered what all the movement was about. Some species, like salmon, are *anadromous:* they go to the ocean to feed and return to fresh water to spawn. Others, like some eels, are *catadromous:* they spawn in the salt and go to the fresh to feed. All have had to develop sophisticated biochemical mechanisms and extraordinary homing abilities.

A recent report in *Science* magazine stood back far enough to take in the whole crosshatch of migrations and come to a conclusion that is startlingly appealing: Anadromy and catadromy show a global pattern. Fish that spawn in temperate-zone rivers move to the oceans to feed, and fish that spawn in the tropical ocean move to tropical rivers to feed. In both cases, the fish are moving from poorer to richer feeding grounds: salmon rivers contain less food than northern oceans, and tropical oceans contain less food than the soupy equatorial rivers.

Like most of our "rules" about nature, this one doesn't always hold. Some rivers host both anadromous and catadromous fish; a batch of other factors come into play in these situations, and the tangle of circumstances affecting the migrations gives rise to many intriguing questions.

But you have to be careful with these questions. You can stand back so far that the only answers are too generalized to yield much satisfaction. If you simply ask, "Why do the fish go at all?" or "Why do some fish go and others don't?" you may quickly get lost in a morass of evolutionary possibilities. Faster growth (as happens in sea-feeding fish) may result in higher survival rates or greater reproductive capacity; a population of fish that is spread over many watersheds (as occurs as stray salmon colonize new streams) will for reasons of numbers and diversity be a population better able to survive local catastrophes.

On the other hand, a tiny fish that lives in a few square yards of river bed may already have moved too far in one evolutionary direction to have success going in another. These sorts of generalizations really tell us little more than that, in any given ecological situation, nature will exercise all sorts of options at once.

Evolutionary directions taken by species aren't the result of conscious decisions. No one ever promised salmon that in the long run becoming anadromous would be just the thing. They evolved in that direction for no more reason than that their habitat, physiology, and appetite permitted them to. We can assume they're still evolving. Indeed, there are a number of fish species in North America that live primarily in rivers but occasionally run down to the tidewater for a few meals, kind of doodling around in the brine for a while like they're waiting for the next excursion boat to Vancouver. (One imagines a pair of them, a male and female in a Gary Larsonesque cartoon, standing around discussing it. She says, "All the neighbors are becoming anadromous, but do you ever take *me* to Alaska? Noooo.")

So perhaps the question is, Where is all this traveling getting the salmon, besides right back where they started? Are they going somewhere in particular, evolutionarily? Or, to ask the question from the popular bottom-line perspective, Was anadromy a good idea?

Anybody who has followed modern fisheries management knows that andromy's gotten them into a lot of trouble with humans. Periodically gathering your whole population up and cramming it into small exposed spaces is only a good way to make it as a wild animal if you're prepared to take some pretty heavy losses to predators (salmon are a piscine version of the biological storm, or at least they were once). Pacific salmon thrived for thousands of years despite many predators, including native Americans, but European technology and greed were too much for them. Long before we developed ways to decimate oceanic fish populations, we were hauling salmon out of streams in disastrous numbers.

So the answer may be, yes, anadromy works pretty well, as long as you don't encounter some fabulously overevolved predator, that once-in-a-billion-year freak (let's face it, we're not your average animal) that is the average species' equivalent of an Act of God.

Salmon have made a magnificent adjustment to the food-poor rivers where they spawn; when they return from the ocean, they eat virtually nothing while in fresh water (this is also true of Atlantic salmon and steelhead trout, two species that live to spawn repeatedly). This grand mass of animal life sweeps into the rivers with a single appetite, one that makes no nutrient demands on the river's biological resources, but contributes to them enormously.

When tons of fish come up from the sea, many a wild

creature gets fed, from mink to bears to eagles. Some feed in the water, some haul salmon a hundred yards from the stream to be eaten at their leisure. One recent study, conducted by the Washington Department of Natural Resources and the National Park Service on the Olympic Peninsula, had the happy title of "Ecology of Dead Salmon." The researchers tagged nearly a thousand carcasses and installed tiny radio transmitters in 174 more, on four streams, in order to track just what became of all that fish meat. They described the fate of a typical coho salmon carcass:

> A sequence commonly observed was for raccoons, otters, or bears to retrieve carcasses from the stream (bears removed fish from pools more than 1.5 meters deep), to feed, and then to move on. The scattered remains were next consumed over several weeks by small mammals and birds, until only scattered bones and pyloric caeca remained.

These leftovers, the report went on, frequently showed the tiny teethmarks of shrews and rodents, suggesting that eventually every piece of salmon would be consumed, digested, and then distributed on the forest floor. To all of those salmon eaters, yes, anadromy was a good idea.

That study and others suggest that a lot of salmon meat isn't consumed by birds and land mammals. If the river is too deep or too swift, or for some other reason has stretches inaccessible to predators and scavengers, the dead fish remain in the stream, where they decay and may have a substantial effect on the basic organic character of the stream. As gardeners know, fish make great fertilizer; imagine what a few thousand twenty-five-pound salmon would do for the aquatic vegetative growth in a mile of river. In the organic fortunes of the river, anadromy must be judged a terrific idea.

With this information in mind, you can stand right up close to the river and see what an awful tragedy it is that the great salmon runs that once filled it are mostly gone. But stand back a little farther and you'll get an even better idea of what the tragedy means; it's not just the loss of a beautiful spectacle, or of good fishing. The spawning run isn't an event; it's a process. When those millions of carcasses are no longer feeding all the creatures that lived along the stream, the scavenger and predator populations decline or disappear. When carcasses no longer settle to the bottom or get hung up in streambed tangles, the stream's nutrient load declines, and its capacity for supporting other life forms does too.

To make matters worse, humanity's ambitious labors at "improvement" of navigational channels, and prettifying rivers in general, have resulted in the removal of all the stumps, snags, beaver workings, and other obstructions that not only provide shelter for living fish but also hold carcasses in place after the fish die. Many rivers that were once heavily laced with natural nutrient traps now give water a straight shot to the sea, and flush out clean in a good freshet.

So I finally stand back far enough to come full circle and, like the salmon, return to the shallow gravel bar where we started, one haggard, frayed old hen salmon and one puzzled, awed fisherman. The odds against her getting home were great enough before we started messing around with her river; now it seems amazing that any of her kind show up at all, and sad that she won't even be able to die and rot the way she once would have. My sadness isn't the result of some anthropomorphic concern that she be allowed to die a dignified, natural death—I couldn't begin to define such a thing. My sadness is that she represents something badly and ignorantly damaged. For all my resistance to anthropomorphism, it seems to me that as I watch her swim listlessly up the gravel bar to a quiet resting place, an apology might be more in order than a welcome.

PART III

BY ANY OTHER NAME

JIM HAYS

Taming the Giant Weasels

Near the conclusion of *Romancing the Stone*, a marvelous escapist adventure movie, the Arch-Bad Guy crouched on the shore of a tropical harbor, triumphantly clutching the priceless gem that had absorbed the attention of the entire cast for nearly two hours. Suddenly a large alligator poked its head out of the water and neatly bit off the hand holding the gem. In that moment, a great deal of justice was done, and good soon triumphed.

Now this isn't what an alligator would have done; it would have grabbed the arm and pulled the man under to drown him, then consumed him at its leisure (the average human hand has very little food value; the average gem has none). Alligators don't have the sort of teeth that would allow one to neatly snip through an arm. The alligator would have removed the arm, if it wanted to, by twisting its head around until the combination of torque, tearing, and pulling brought the arm free.

I don't introduce this bit of cinematic excess for the joy of discussing the gruesome. I find it revealing in more important ways, because it says a lot about our willingness to ascribe remarkable powers to animals we routinely consider ferocious.

Ferocity in nature is a cherished social concept. Wildernesses howl, beasts are savage, and anything—oar, hand, foot—left exposed over the edge of anything else—boat, bridge, shoreline—is liable to disappear into a ferocious mouth. As children we are careful not to let a foot or hand dangle over the edge of the bed, for fear of the ferocious menagerie of carnivores (mine were for some inexplicable reason giant weasels) that roam the floor after the lights are out. Vicious bears, vindictive rattlesnakes, and simply vile crocodilians stalk the pages of our favorite outdoor magazines, filling readers with ecstatic loathings. This may be our fullest love-hate relationship.

I don't for a minute question the reality of what wild animals do to each other, much less what they occasionally do to people. Almost anything alive in the woods today might be eaten by something else tomorrow (or tonight, darkness seeming to be when ferocity is most often rampant). What I question, in fact what I am inclined to discount, is this whole business of describing some animals as ferocious in the first place.

Webster tells me that ferocious means "having or exhibiting ferocity, cruelty, savagery, etc.; violently cruel, as *ferocious* actions or looks." That seems clear enough. It describes some types of human behavior—war, for instance—quite well. But does it really serve as an accurate description of other animals?

There are a few animals we routinely consider ferocious: bears, wolverines, cougars, and a host of exotic predators. They are all carnivores, having evolved to survive by eating other animals. If by "ferocious" we mean "can be expected to kill and eat other animals," then these animals are surely ferocious. But if by "ferocious" we mean "is characterized by a need to cause pain and be cruel to other animals," then we're on thin behavioral ice.

I don't doubt that some—maybe many—predators feel an elemental satisfaction or exhilaration when bringing down prey. I do doubt our implying that the process of killing their food is either motivated or encouraged by some need to be "violently cruel." These animals have evolved to get a certain job done efficiently, an evolution that is as much a response to the traits of their prey as it is to the needs of the predator. Killing is a biological function first, whatever else it may be afterwards. A bear uses pretty much the same shearing-tearing motion when cropping grass as when separating meat from a carcass. Rarely is there any evolutionary purpose in causing physical pain beyond that necessary to acquire food. A lynx is not encumbered by human concepts of good sportsmanship, but it has every reason to make a clean kill.

It gets complicated when a predator eats its prey without killing it, but neither the predator nor the prey knows our rules of good sport, and all the predator really must do is immobilize the prey. Sometimes that means the prey is dead, sometimes that means it's not. Besides that, predators have been observed "playing" with captured prey before killing it, or engaging in what even the ecologists regard as "excess" killing, taking more animals than they can immediately eat. These activities, however well they may work for the predator, are hard to watch without emotional reaction, but on the whole the process by which carnivores meet their nutritional needs, though it may seem untidy to us, is not nearly as "ferocious" as the acts of many human cultures.

What I suggest is that we are overcome by empathy. Ferocity is a characteristic we ascribe almost exclusively to those animals that are hurting animals we sympathize with. Our conception of ferocity has as much to do with who is being eaten as with who is doing the eating. We may be shocked to see a deer that has been killed by a predator, but we enjoy watching a swallow wheeling about over a pond, consuming hundreds of newly emerged midges. The swallow—a lovely creature identified with love—has just done violence to hundreds of fellow creatures, but they are not creatures we get very excited about. Is the swallow's behavior any less aggressive, or ferocious, than the behavior of the cougar that killed the deer? No. If anything, the swallow is a more effective predator, killing by the hundreds, daily. It's just that we don't have any sympathy for the swallow's prey.

Here is a good test of how we perceive predation, one I suggested in my book *The Bears of Yellowstone* when I was trying to get readers to reconsider the legendary ferocity of the grizzly bear. The next time you disturb a blue jay, and he is raising a mighty ruckus in the branches right above your head, imagine what he would do to you if he weighed about 600 pounds. That's about as ferocious as could be, and surely, right now, weighing only a few ounces, that blue jay is just as upset as he would be if he were ten feet tall. The difference is size and ability to do harm, not intensity of belligerence. Ferocity, at least as we are inclined to define it, comes down to how threatened we are by the animal in question.

In the past couple of centuries, we've gone through a variety of attitudes about animal personalities. At some periods a purely mechanistic outlook has prevailed, wherein all animals were considered fundamentally thoughtless automatons reacting mindlessly to their surroundings. At other periods, we've been on binges of anthropomorphism, equating animal actions more or less precisely with corresponding human actions. Most of the time, no matter which view the majority adopted, we were weighing down the animals with our own preconceptions and moral positioning. In his excellent history of human-animal relationships, *Man and the*

Natural World (1983), Keith Thomas describes the sixteenth-century British in a way that often still applies today:

> Men attributed to animals the natural impulses they most feared in themselves—ferocity, gluttony, sexuality—even though it was men, not beasts, who made war on their own species, ate more than was good for them and were sexually active all the year round. It was as a comment on *human* nature that the concept of "animality" was devised. As S. T. Coleridge would observe, to call human vices "bestial" was to libel the animals.

I don't mean to suggest that we haven't made progress since the 1500s. Gradually, I think, we've reached a more realistic balance in how we view animals. But most of us still carry a pretty heavy load of value judgments, and there is probably little harm in that. I am personally certain, for example, that the obnoxious dog that lives across the alley is not simply a poorly trained dog; he is a jerk.

The popular view of predators as mean-spirited villains, pervasive as it still is in children's literature, cinema, and urban adult attitudes, is unfortunate. I for one would have been happier if, at the end of that entertaining adventure movie I mentioned earlier, the alligator had behaved realistically and dragged the bad guy under whole, providing an opportunity for a most dramatic cinematic death, the churning tropical pool that becomes ominously still. It certainly would have been a fitting end for the ferocious villain.

Mirror, Mirror

When we lived in south Texas in the mid-1950s, my family went for a picnic along the Nueces River. My mother and sister didn't care much for fishing, but my father and my older brother did, and I was just interested enough that they took me with them. Steve was about twelve, and I was seven or so.

I don't think we caught anything from the Nueces, but of all the places that we fished when I was small, the Nueces was the one that most formed my idea of what fishing could be—not a way to gather food, or have fun, or get exercise, but a sort of quest. The Nueces, unlike everyplace else we fished, where really big fish were just an idle dream or something we somehow knew only others would catch, had fish so huge that I was genuinely frightened to get close to the water.

We saw evidence of them along the shore. Fireplaces and picnic spots were littered with big, shiny, flat things—in my memory, they were the size of Chevy hubcaps—that I was stunned to discover were scales. I knew only the dainty little scales of bluegills, and I remember not fully believing that there could be a fish big enough to need more than one or two of these dinner plates per side. I had no idea.

We fished at a deep, wide, still stretch of river. I don't

132

remember fishing at all, though I must have. I do remember my father—a broad, strong man—standing right along the bank, lobbing a big hunk of weighted shrimp or some other meat out into the current with his heavy old casting rod. I remember the shore was dusty, and like everyplace else in Texas, it was hot.

Mostly, though, I remember the gar. Alligator gar have grown to more than 300 pounds, more than nine feet long. I don't suppose the ones we saw were that big, but even allowing for the amplification of memory, they must have been four or five feet. Let's say five. Maybe six. Maybe I'd better admit it: I still think they were at least ten. Out in the middle of the river, about as far as my father could cast, one would roll every now and then, a big, slow, churning turn on the surface, baring its dully glinting back for a moment to the hot Texas sun.

Even at that age, I had seen big fish; from the wharves along the coast my father had pointed out the moving fins of tarpon and sharks, and I'd watched big hammerheads taken from the pier at Padre Island. But that was the ocean, where, given enough horizon, nothing looks too big. In the Nueces, at that place, at that age, and at that distance, the gar was more a monster than a fish, and my clearest memory of all is of the moment following the fish's roll, when my dad would put his considerable muscle into a cast aimed right at the still-swirling spot where the fish had just surfaced. I was scared beyond words that he would hook the fish and it would pull him in.

Here was fishing with proof for the effort. When those big fish rose and pushed against the boundary between their world and mine, they very nearly fulfilled the fisherman's dream, merely by satisfying the hope we all have that there really is a chance, however faint, of catching a monster.

Here also was a different kind of promise, one I only grew to believe fully years later, when I learned that the rolling of the gar was only one of countless ways in which the surface of the water could tell me what was beneath it; that water was more than a mirror to reflect its surroundings, it was the source of immense curiosity and entertainment.

Memory of the gar came back to me recently along a small western backcountry stream. I was introducing two friends to some of the techniques of fly-fishing, making a few casts to show them where to find fish and what to do about it. I stood at a long, still pool, where the water was a little murky and the far bank was thirty or forty feet away. Counseling my companions to watch closely, I cast across the stream at a downstream angle, so that the large fly plopped noisily into the shallows along the far shore. I then began a series of quick, jerky retrieves, pulling the line through the guides and the fly back toward me, across the deep pool.

When the fly was about a third of the way back to me, a good-sized trout swirled to the surface behind it, lunged at it, and turned back into the deep water. With some excitement in my voice (nothing makes a fishing lesson more instructive than a fish), I said, "See that?"

Both my friends stared blankly at the water. "What?"

They had been looking at the water; they had been following the progress of the fly with considerable interest. But neither had known what they were looking for. The swirl of the fish, into which I had read much meaning, had not even registered.

I, on the other hand, having spent a few thousand hours looking at the surfaces of lakes and streams since that day on the Nueces, had grown to assume that this stuff was visible to everyone. Jolted back into reality, and a little surprised at how completely I'd lost touch with the nonspecialist's perspective, I made a poor attempt at explaining how a fish will disturb the surface when it moves, whether it moves for food or flight or other reasons.

Fishermen have been reading these surface disturbances for a long time. By 1930, when Eric Taverner published his comprehensive *Trout Fishing from All Angles,* there was a whole subdiscipline of fishing devoted to the analysis of what are still called "rise forms." During a heavy emergence of aquatic insects, the smooth pool of a trout stream may be pocked with these little circles, places where the trout broke the surface to take their prey. Rise forms are like rapidly fading tracks: the fish rises, a bulging ring appears on the surface, the fish drops back down from the surface, and the ring is carried away with the current, healing into smooth water in a few seconds. Taverner listed more than a dozen types of rises, describing not only their appearance but their cause.

Their cause, of course, was the reason for caring at all. This was not an abstract exercise in aquatic aesthetics; a smart fisherman could tell a great deal about the trout's behavior by reading the rise.

For example, fish sometimes feed just a little ways beneath the surface, preying on the immature forms of mayflies, caddisflies, and other aquatic insects as they rise to the surface to emerge into adults. At other times, they feed on adult insects that drift along the surface drying their wings. Trout often get very choosy about what they will eat, and lock in on one particular food form to the exclusion of all others. The fisherman wanting to know whether to fish a floating fly or a sunken fly must be able to distinguish between the respective surface disturbances created by these different feeding levels, and rise forms give a great deal away.

A fish that is actually breaking through the surface film to eat a floating insect will almost unavoidably inhale some air with the insect. Once the insect has been taken, the fish settles back to its resting position to watch for more food. But the air it took in with the insect is not swallowed; it is expelled through the gills and returns to the surface, where it appears as bubbles in the middle of the rise form. Thus, any rise form accompanied by bubbles indicates a fish feeding on the water surface.

Besides revealing where the fish is feeding, the rise form may reveal the type of insect the fish is eating. As Taverner put it, "The shape of a rise form when the surface of the water is broken is the measure of a trout's anxiety to annex a fly, an anxiety caused by fondness for a particular insect, by the size of the insect, or most frequently by the fear of losing it." Trout and bass feed most visibly—or violently—on prey that seem to have a chance of escaping. Trout are known, for example, for their splashy rises to grasshoppers that have fallen into the water and are vigorously thrashing the surface in their attempts to get out. Trout recognize a rare and short-lived opportunity, and sometimes feed on grasshoppers with great energy and abandon, inspiring fishermen to develop all sorts of imitation grasshoppers.

Bass feed on the surface with even greater abandon. Largemouth bass will inhale almost any living thing that seems even remotely edible—snakes, ducklings, smaller fish, frogs, salamanders—and do so with such explosive effects on the water's surface that fishing "topwater" lures is a great favorite among anglers. No matter how often you have done it, you can't prepare yourself emotionally for the sudden attack of a big bass on your lure. Professional fishing guides I know tell me that more than one fisherman has wet his pants under these circumstances.

It has proved a hard sell among the general public, but several naturalists I know have promoted the idea that fish-watching is sometimes as rewarding as bird-watching. Most people, even those interested in nature, tend to neglect the appreciation of aquatic environments. After all this time of sloshing around in the water, I have perhaps reversed that bias; I now tend to focus on the water in a landscape, often to the exclusion of the hills and meadows.

I do this in places where I have no intention of fishing. There is a special pleasure in exercising this hard-won vision, and I recommend it. I recommend it even if you cannot see the fish. If the water is too murky, or the light is all wrong—a heavy overcast can turn even a clear trout stream a leaden gray—you can still read the water.

Evening is perhaps the best time. Go to whatever water is nearby; a farm pond is a good place to start. Get your nose right down to the water and look for signs of life. The most obvious will be the invertebrates, all the little backswimmers, water striders, and assorted nymphs that stir the surface with their wanderings.

But be prepared for bigger game. Find a vantage point that gives you a good view of the pond's various ecological borders.

The places where the water becomes deep enough for large fish are often the places where those fish can find the most food; careless minnows drift beyond the safety of the shallows, and the sun-warmed shoals are very productive of the insect life that fish relish.

The edges of reed beds, whether near the shore or not, provide cover for both predator and prey. Larger fish will sometimes lie in wait, backed into heavy weeds, until some unwary smaller fish comes by. All you may see on the surface is the occasional peculiar movement of the top of a reed, as the big fish maneuvers through the tangle underneath.

Most of all, learn to look at the surface. Even on windless days, it is never completely still. As the evening passes, you'll see it peppered with raindrop-size rings as aquatic midges emerge from below to mate in the air over the pond. You may see tiny swirls near the shore as a dragonfly nymph, one of nature's most powerful predators, captures some smaller insect.

But with any luck at all, you'll see at least a few bigger tracks, as a bluegill gently takes down some anonymous

137

bug with what the fishing writers aptly call a "kissing rise," or as a small bass rolls up under a minnow. And if you keep at it long enough, and practice the patient humility only a few fishermen ever master, you may witness some greater disturbance, some rushing breach in the water so large and violent and unexpected that you too will suddenly, and always thereafter, believe in monsters.

A Lot of Browns and Yellows

On June 13, 1987, the U.S. Postal Service issued a handsome set of fifty stamps to "commemorate the natural heritage of North America." The stamps, executed in color by prominent wildlife artist Chuck Ripper, represent a cross section of popular animals from all fifty states. Ripper is prominent because he is skilled, and the animal portraits, considering that they have to fit into a format whose very name has come to mean small and difficult to see (as in "When he went into his batting crouch he had a strike zone about the size of a postage stamp"), are beautifully done. And yet these innocent and attractive little portraits are likely to warm the hearts of natural history polemicists, for they appear at first glance to reinforce some troublesome stereotypes.

I can imagine several objections that might surface among nature lovers. The most obvious one involves the omission of personal favorites: "How could they exclude the eastern bluebird?" or "Why is the box turtle more important than the American crocodile?" or "Why put a varmint like the woodchuck on there and leave off the noble caribou?" Arguments about this sort of thing tend to mire down in subjectivity; this one is more or less settled by recognizing that once

139

it was decided to include only fifty it was simultaneously understood that no fifty would satisfy everybody.

There are some tougher questions, however. Why, for example, are thirty-one of the fifty species depicted on the stamps mammals? Only thirteen are birds, four are insects, and one each is a reptile (box turtle) and a crustacean (lobster). Does this suggest that selections were based on the traditional notion of "wildlife" as warm, furry things with pretty eyes (the four insects—two butterflies, a moth, and a ladybug—all have pretty wings instead). In the natural world there are many times as many insect species as there are mammal species; if we can't have proportional representation, couldn't we at least expect a little better balance?

There are no fish included at all, but then most people don't even think of fish as wildlife in the first place. Fish live in water, and most people's definition of wildlife is still limited to terrestrial creatures. But why no snakes or amphibians? Are we still slaves to the old "glamor animal" mentality that for so long ruled state fish and game agencies, the mentality that measures an animal's worth to the world only in terms of its ability to create sensation or its simple economic value? Does it have to sell hunting licenses, earn huge foundation research grants, or look good on a T-shirt or a calendar to deserve our affection?

But even these questions, though a little harder to ignore than the ones about leaving out this or that favorite animal, are not an entirely fair challenge to the stamp committee's chosen fifty subjects. What we really should be asking when we challenge the choices is, "How did this selection happen?" And, if we are open-minded enough, "Is it possible that my disappointment is the result of my own lack of understanding?"

D. B. Rustin, manager of the Stamp Information Branch of the U.S. Postal Service, explains the mechanics of the selection process in terms considerably broader than mere fair representation. This isn't, he observed, just a matter of depicting important animals; it is a matter of creating a pleasant, reasonably well balanced collection of colors on a big sheet of fifty different pictures. The color balance may mean nothing to the naturalist (or it may mean a lot but for different reasons), but it means a great deal to the people who must produce stamps that will sell, and who know, in the words of Mr. Rustin, that "the prettier they are, the more pleasing they are to the customer."

Pretty colors are a big factor in all stamps, whatever the subject; wildlife is merely being squeezed into the same commercial mold as traditional American crafts, famous poets, and secular Christmas greetings. You want your enthusiasm immortalized on stamps, you have to pay the price.

This doesn't let the Postal Service entirely off the hook for the criticisms mentioned earlier. And it doesn't let the public off the hook for still requiring such things as prettiness and big cow eyes in its wild animals. But it does go a long way toward explaining why twelve of the thirteen birds are strikingly colored (scarlet tanager, roseate spoonbill, snowy egret); the thirteenth, a mockingbird, thereby becomes distinctive and even striking just because it is uniquely plain in the group.

It also explains, according to Mr. Rustin, why the birds are there in the numbers they are: "If you just use mammals, you get a lot of browns and yellows. We put things like ladybugs on to create color." If color is indeed necessary to sell the stamps, then the task of getting fair representation for camouflaged, stinging, squirmy creatures is going to be a tough assignment. Making people understand the ecological value of alligators is one thing; making them recognize the beauty of a monochromatically tinted caddisfly is another.

For all their easily criticized limitations, the stamps actually contain many hopeful signs. The inclusion of several mammals that have traditionally been thought of as "bad" (wolverine, wolf, mountain lion) is a heartening indication of the substantial progress we've made in appreciating the role of various elements of the mammalian food chain. Eleven of the thirteen bird species fall under the "nongame" (which to wildlife managers traditionally meant "nonprofit") heading so popular in modern wildlife parlance, and though several of them are "cause" animals with their own constituencies and romantic associations (bald eagle, osprey, egret), several others are not.

There's no pretending that the four insects are representative of anything except photography ads, but it's worth something that there are four insects at all. The lobster is about as mud-colored and noncuddly as an animal could be (the Postal Service resisted showing the rare blue phase). And, best of all, not one of the predators is snarling or acting the way they always act on the covers of outdoor magazines (though the wolverine does look a little peeved about something).

So even if the selection of these lovely little stamps is judged by some to be inadequate, or biased, or even wrong, it is a promising selection anyway. If you would rather have had a manatee than a monarch, or an arachnid than a canid, that too is promising; the very fact that the list can evoke our disapproval is a sign of progress. Fifty years ago, would many people have cared at all?

* * *

Shortly after the preceding piece appeared in *Country Journal*, I received a letter from Nestle J. Frobish, who introduced himself as "chaircreature" of the Worldwide Fairplay For Frogs Committee (Box 94, Lyndonville, VT 05851). After a little correspondence, I realized that Mr. Frobish was serious, in his own lighthearted way, so I now want to share with you two letters from him. The first was written to me:

Dear Mr. Schullery:

Please accept my compliments on your brilliantly perceptive [How could I dislike a guy with such penetrating judgment? *P. S.*] article in the November 1987 *CJ*, on the subject of wildlife stamps.

As frog partisans, we have already had a conniption about this episode. I enclose the correspondence to date.

"But why no snakes or amphibians?" indeed.

Whatever explanation you may have received, I am convinced that this is just one more power play by the bird-and-mammal boys. Someday they'll get theirs.

Keep up the good work.

Yours Truly,
Nestle J. Frobish, Chaircreature

And here is Mr. Frobish's first letter to Mr. Preston Tisch, Postmaster General of the United States Postal Service:

Dear Mr. Postmaster General:

I regret that it is necessary for me to call your attention to a fairly serious problem, which may well have originated inadvertently. At least we hope so.

Your Postal Service recently issued a fifty-stamp sheet set of American wildlife, handsomely portrayed in full color. If you will look over that sheet, you will find that it includes thirteen birds, four insects, thirty mammals, one crustacean, one reptile, and one armadillo, whatever the hell that is. From this recitation the horrendous thought should come to you: you have left out amphibians.

How is it, Mr. Postmaster General, that you can portray not one but two each of bears and rabbits; plus some bird with a name which hitherto has been regarded only as an asthmatic wheeze; plus a lobster, a species which in all recorded history has exhibited no intelligent or endearing qualities other than its apparent willingness to be boiled alive; plus, for Chrissake, an armadillo, whose only claim to fame is being squashed by eighteen-wheelers on highways all over west Texas?

Perhaps the dim bulbs down in your stamp selection bureau have forgotten the long and patriotic record of American FROGS. When General George Washington plodded through the Jersey swamps, he was encouraged by the frog chorus croaking "Rip Brits!" The same encouraging sound greeted Andrew Jackson and Jean Laffite when they whipped the British at New Orleans. Frogs greeted the first 49ers in the gold streams of the Sierra, and cheered on Lewis and Clark on their epochal journey to the Pacific. I could go on, but I think you get the idea.

Mr. Postmaster General, on behalf of millions of American frog lovers, we of the WFPFFC firmly request that you make up for this calamitous oversight by putting out a commemorative stamp honoring the humble frog, friend of mankind whose pleasant *chuggarumm* has improved so many summer evenings throughout this great land of ours. Nothing against birds and mammals and insects, mind you, but now they've had theirs. There are certainly fifty highly attractive species of frog that could make up your new stamp sheet; or, to fend off some potential discontent from other quarters, you might want to include a few newts, toads, salamanders, efts, and skinks. We're not greedy—but don't overdo it.

Please let me know promptly when we can expect this new sheet to appear.

Yours Truly,

Nestle J. Frobish

Mr. Frobish may be a little too hard on armadillos and lobsters, but what with the worldwide decline in frogs and toads that is making the news these days, I wish him the best. I'd kind of like to see a commemorative eft stamp myself.

So Long, Sucker

As the wildlife stamps, for all their limitations, suggest, unattractive animals have it a lot better than they used to. It's been a hard fight, but now that we're living in the Enlightened Present, we've learned that it's reasonable and sometimes even fashionable to approve of animals that used to give us the willies. Spiders, bats, and snakes get equal billing in wildlife magazines with cuddly bears and spectacular birds. Wolves, hated for thousands of years, now are not only OK but are downright nice. To us, this all seems like a great thing. It makes us feel good to be so open-minded, and maybe it should. Of course, to the animals (if we could somehow ask and they could somehow answer) it must seem a little late in coming.

But this business of judging animals is not a simple matter of saying, "OK, now we won't mind that the wolf eats baby moose, or that spiders make our skin crawl." We can go through the intellectual exercise of changing our collective mind, and we can back up the change with new laws and new educational programs that will do much to even out public attitudes. But revulsion runs deep, and anything that took centuries to construct isn't going to be knocked down by a few magazine covers and movies. Some of the animal world's second-class citizens are set up for a lot more disappointment.

Our language is still full of cheap shots at innocent life forms, shots that suggest how hard it will be to really change our attitudes. A man who *wolfs* down his food is a *pig*. A treacherous person is a slimy *snake* (real snakes aren't even slimy, so this guy must be awful). "Quit *sponging* off me and get a job, you *leech.*" We have entangled human traits and wild animal behavior beyond all separation, and I don't think all of them can be sorted out; the names are just too loaded now. For some animals there's probably no escape, and I suspect that the most hopeless prisoner of human language is the sucker—who, it is clear, is never going to get an even break.

The sucker is the real "boy named Sue" of the animal world. What a word to name an animal. If all the intriguing new concepts you learned about this word in junior high school aren't proof enough of how unkind it is, take a look at the nineteen suction-related columns of fine print in the Oxford English Dictionary. There you will find a thousand years of collected condescension and insult, such breathtakingly imaginative uses of the root word "suck" as would stun into silence the most foul-mouthed eighth-grader.

But we didn't stop there. Calling this pleasant little fish a sucker wasn't enough. As species were recognized, they got a second round of insults. There are many kinds of suckers, among them hog suckers, carp suckers, short-nosed suckers (getting pretty personal), yellow sucker (getting even more personal), flannel-mouth sucker (getting abstrusely personal), and chub sucker. Some of those names would have been pretty valuable in junior high school too.

Worst of all, the name isn't even accurate. In the modern vernacular, a lot of things suck, including vacuum cleaners, drains, babies, final exams, and many situations (as in "This job really..." or "This weather really..."). But suckers, the fish, don't. At least, they don't any more than other fish do.

Look a fish in the face, or, better, watch one in the water. A fish breathes by letting water in through its mouth and, rather than swallowing it, sending it out through the gill openings on either side of its head. The water rushes over the gills, which extract oxygen from it.

Feeding is accomplished by the same procedure. Food is drawn into the mouth with water, but is captured in the mouth or throat while the water flows out through the gill openings. Most fish, including suckers, operate on some variation of this procedure. Sometimes the process is as simple as the fish keeping stationary in a current and letting the force of the current push the water through its face. Sometimes the fish may generate some of the force itself, pulling in some water by working its gills, perhaps even creating a little "suction" in the process. But it's nothing like what a vacuum cleaner does.

The white sucker, one of the most common suckers in North America (and, by the way, the majority of the world's suckers do live in North America), starts life as a regular-looking little fish. When it's a week or so old, its mouth is out on the front of its face. The eyes are directly behind the mouth. It's a very straightforward configuration of facial features, but it doesn't last.

The tiny fish, having absorbed its yolk in the first week, then feeds near the surface of the water, eating tiny free-floating organisms just as a newly hatched trout would. This lasts for about ten days. Then there's a period just about as long, during which the mouth begins to droop and the eyes start to move upward; the fish shows more interest in stuff on the bottom of the pond or stream. At the same time, the little sucker (notice how hard it is to read those words without loading them, as in "Gotcha now, ya little sucker") changes its habits, and identifies depth and bottom cover with safety as well as with food.

By the time it's twenty-five days old, the sucker is a bottom feeder. Its mouth has moved to the underside of its "chin," and the eyes are up closer to the top of its head. It feeds by moving along the bottom, nudging and dislodging algae and insect larvae and nymphs. As it gets older, it is better and better able to sort all the stuff it takes into its mouth, ejecting the sand and assorted nonnutritive crud that got there by mistake.

This all means that suckers are frequently harder to see than sport fish that feed on the surface, even in waters where the suckers outnumber the sport fish. What people passing by most often see, if they see anything at all, is the slow flash of the sucker's side as it rounds the contours of a boulder or tips over to root in some silt. If the whole fish is visible and you can watch it work, it looks for all the world like it's vacuuming up the river bottom. Those tough, bottom-oriented lips are perfect for sorting through the plant film and muck.

I think the lips have had as much to do with the sucker's name as has its vacuum cleaner-like behavior. A sucker's mouth looks enough like the "suckers" on an octopus's tentacles, or like the mouth of a leech, to trick casual observers into assuming that it operates in the same way, that is, by attaching firmly to something and holding on tight. But someone could hold a sucker's mouth against your arm all day (if you were of a mind to let them) without it ever once taking hold. It doesn't operate on suction, but it kind of looks like it does.

As far as its public image goes, the lips and mouth do just as much harm as its supposed sucking behavior. It looks like a person with a weak chin, a blunt oversized nose, and peculiarly upturned eyes that give it a supplicating, self-demeaning appearance. All of those are human impressions, of course, but they can be a powerful influence on our judgment. Especially when the fish in question is also a "foul feeder," rooting through scum for its unappetizing foods.

Among ecologists, the sucker at least has been recognized for its utility in the natural system. There is little reason to believe that in most situations the sucker competes with trout or bass for food; in fact, suckers may help feed the preferred sport fish species. Suckers are "primary converters"; that is, they, like aquatic insects, consume plant matter and turn it into something other fish can eat.

Among fisheries managers, most talk of suckers is of this sort. Suckers, like an assortment of other nonsportfish, are judged for what they may do for the sport fishery. Do they compete with trout for space or spawning grounds? Do they get eaten in good enough numbers to make the trout bigger? Do they stir up the bottom in their foraging, thus releasing more nutrients into the water? Do they stir up the bottom so much that water clarity is affected, thus reducing the amount of sunlight that can reach plant life? Do they become "useless" by growing too big for the trout to eat? In short, what do they do to help?

In this practical view, they at least have some importance and are recognized as real live animals living in complex relationships with other animals. Though such backhanded importance may not be a great thing if you're a trout with pretty spots and a Hollywood image, it's about the best the sucker can hope for. But it has its drawbacks. It means the sucker is still expendable if it doesn't do quite what we want it to do for our preferred fish. John Livingston, a prominent and outspoken Canadian nature writer, talked about this problem in a book published in 1966.

> The word "conservation" means many things to many people. It is used frequently as a synonym for "wise use"—though we are not altogether sure what *that* means, either. The word seems to con-

vey whatever the user wants it to. "Conservation" is commonly used by the forest industry, for example, to describe its minimal gestures toward the restoration of the resources it has so sadly mutilated—especially on the Pacific coast. It is almost as commonly used by organized fishing and shooting groups to justify such activities as the poisoning of water systems so that a pure culture of some exotic game fish may replace the diverse and wonderful natural community of fishes that developed there.

Fortunately, organized sporting groups are becoming attuned to that kind of abuse, but mostly because managers have made them realize that suckers are sometimes "useful." Suckers still too often fall into that category of admiration-exempt creatures known as "trash fish," "forage fish," or, in a powerful colloquialism I once heard from a Vermont fisherman, "shitfish."

It would be nice if, someday, a dad taking his son fishing would happily point out a passing sucker and say, "Look there, Buford, it's a sucker! What a grand fish, playing a critical role in the population dynamics of our bass and trout!"

It would be nice, but of course little Buford's in junior high, and he knows better. "Are you nuts, Dad? It's just a scummy old sucker. Let's kill it for bait."

Microdragons

I saw my first horned toad about thirty-five years ago, and the meeting was not a disappointment. I was in second grade, and we had just moved from Pennsylvania to what was then suburban Corpus Christi, Texas. I was walking across a vacant lot on an errand for my mother when suddenly (as most things happen when you're seven) there it sat, in the middle of the path.

We considered one another briefly, the child and the adult, while my mind whirled with images of warts, dragons, and a universe of unnamed but thrilling revulsions. The horned toad did not move, so I circled widely into the brush and hurried on my way. It was not there a few minutes later when I returned.

What strikes me now about that encounter is that for all I'd heard about horned toads from my new Texan schoolmates, my most fantastic imaginings didn't live up to the real thing. The horned toad is the old wive's tale incarnate. But it is real; no one could have made up such a singular lie.

To begin with, it isn't even a toad. It's a lizard. Like a few of our other western animals (the buffalo, the antelope), it has acquired its name through common usage overriding technical accuracy. The horned lizards picked up the "toad" label because of their squat bodies and their toadlike ability to take insects from a short distance with an invisibly rapid tongue.

The other half of their name is only sort of correct. The many bumps, ridges, scales, and points that adorn the head and body aren't horns in the formal sense, but they sometimes serve similar functions of defense. Few horned lizards seem able to do great harm to any other animal with the horns; their real success as defensive attire is in making the lizard hard to eat.

In his handsome and thorougly engaging little book, *Horned Lizards, Unique Reptiles of Western North America* (1981), Wade C. Sherbrooke, a lizard authority with the American Museum of Natural History, published a photograph of a whipsnake that tried to eat a horned lizard. The snake managed to get the entire head and a bit of the shoulders of the lizard into its mouth and partway down its throat (exactly where a snake's throat ends is an interesting question) before succumbing to the lizard's horns. In the photographs, the horns on the lizard's head can be seen in outline, distending both sides of the snake's skin just behind the snake's head. It's hard to know which animal to feel sympathy for— the lizard, jammed irretrievably into the snake's tight throat, or the snake, barbed and held by the lizard's armament.

There are more than a dozen species of horned lizards (genus *Phrynosoma*) in Central America and western North America, generally found in arid and semiarid country. Most are between five and ten inches long as adults; the one I met that day, almost certainly a Texas horned lizard (*Phrynosoma cornutum*), is perhaps the most common of all, which may explain why it is so popular in pet stores. The species vary dramatically in size, coloration, and placement of horns.

They have the usual complement of lizardly qualities.

They hibernate, some as much as six months each year. They have an exceptional tolerance for hot weather and a corresponding gift for getting along without water. They are remarkably adaptable to environmental colors, and in this quality they begin to live up to the legend. One species, the short horned lizard (*Phrynosoma douglassi*), has been reported in certain desert habitats as being "distinctly pink, with white spots," those colors matching the soil and pebbles, respectively, where it was collected. Another form of the same species, collected in a lava field, was "satiny black, with rich yellow markings—even the gloss of the lava was imitated."

Several species further distinguish themselves, at least from most other lizards, by bearing live young. These usually appear, six to twelve in a litter, in late summer or early fall. The young quickly settle into the horned lizard diet, which favors ants but sometimes strays into other insects and, rarely, vegetation. They develop the skill of hiding from predators by burying themselves in loose soil, something they do by working themselves in through a sort of shuffling series of undulations until only their back or their head is exposed. They usually hide or sit very still rather than flee, because they are not fast runners. What seemed to me that day a standoff (that I lost) was just a little boy misunderstanding a toad that hoped it was invisible.

These are qualities that together may make the horned lizard a little on the unusual side, but certainly do not make it extraordinary among animals. They have other characteristics, however, ones so apparently bizarre that the most creative old wife would wish they weren't true so that she could turn them into tales.

The first one is about half folklore anyway. The lizard has what some have called a third eye. Actually it's not quite an eye, or not a complete eye, or not eye enough to satisfy all definitions. It's called the pineal gland, and it resides on the top of the lizard's head, more or less between its other two eyes.

The pineal gland is covered by a thin, translucent sort of near-lens and has some of the features of a real eye, including a retinalike area to take in light. Though the gland cannot register actual images and transfer them to the brain, it does seem able to register sunlight, or lack of it. It appears, from tests on other kinds of lizards that also have the gland, that its function is to help the lizard regulate its intake of heat. It's also true that science hasn't really figured this one out as well as science would like to, and so it will probably be called a third eye by at least some people for a long time yet. It sure looks like one.

If that isn't an odd enough physiological quirk, at least three species of the lizards have an even odder one. They can squirt blood from their eyes.

This is one of those items of folk knowledge that is actually true, but it is so odd that even scientists, in their technical journals and reports, discuss it with a barely suppressed sense of wonder, as if to say, "I thought you guys were putting me on, but these lizards really do shoot blood out of their eyes!"

Reports of how far the blood travels vary, but distances of three to six feet are reliably documented. Herpetologist Raymond Ditmars, writing in the 1930s, described the process in great detail, explaining how a Mexican species, becoming agitated while Ditmars was measuring it, lifted its head—its eyes seemed to bulge—and sent a stream of blood "as fine as a horsehair" from the closed eyelid. The blood "hit the wall, four feet away, at the same level as that of the reptile." The lizard then refused to open its eyes for several minutes.

Sherbrooke's book offers the best and most current explanation of the physical process. As it happens, the lizard has developed a way of concentrating blood near its eyes. It has done so at least partly as a way of cleaning dirt from its eye, an important skill for such an energetic digger to have. Like many other reptiles and birds, the lizard has a transparent "nictitating membrane" that sweeps across the surface of the eyeball like an extra eyelid. The ability to increase blood pressure around the eyes allows the lizard to expand small blood sinuses around the eye, by which action any dirt particles are pushed from their lodgings in the corner of the eye.

It appears these same muscles act to build pressure for squirting blood. Sherbrooke described the action:

> There are two sets of constricting muscles surrounding the major veins leading from the head. On contraction, one pair of these muscles restricts the exit of blood from the head, causing the blood sinuses or reservoirs in the head to fill .... Because of the location in the circulatory system of this first pair of constricting muscles, some venous blood still leaves the head via minor by-pass veins, even when the major veins are closed. But the second pair of vein-constricting muscles, located closer to the eyes, can, when constricted, completely block the return of venous blood from the region of the eyes and force the blood pressure here to still higher levels. Under this pressure, blood breaks through the walls of the blood sinuses located in the eye sockets. It then flows into the membranes of the eyelids and sprays forth from the pore of a gland on the edge of each lower eyelid.

Unfortunately, the squirting doesn't happen regularly, so it has been difficult for scientists to study it. It's hard to know to what extent the action is involuntary, and, just as important, what its function is.

Offhand, it doesn't sound like all that good an idea for

an animal that is apparently fearing harm. Usually one does not associate flowing blood with defense, but rather with the attraction of predators to the scent of the blood. It has been suggested that certain glands in the lizard's head may secrete some substance into the blood that makes it annoying or unpleasant to predators, and this remains the most likely explanation. A second liquid is proposed, originating near the eyeball, that comes out with the blood, and a number of canids, especially coyotes and foxes, have been seen to react with discomfort when sprayed.

Horned lizards have adapted to their world in some odd, even startling ways. They have also become extreme specialists; some spend most of their lives haunting ant trails and anthills and eating nothing but ants. Because of those traits and what they mean, think twice about making a pet of one of these stoic, slow-moving creatures. Some do thrive, and they're terribly popular with children because it's like having your own microdragon right there at home, but most can't adjust. Because of their hardiness they may seem to be doing just fine for months, but that's only because they starve to death much more slowly than do most other animals.

While a friend and I were visiting some Anasazi ruins in New Mexico, our guide and mutual friend Chris Judson, a naturalist at Bandelier National Monument, spotted a young horned lizard hurrying among the grasses and potsherds. She picked it up and showed us how children turn the animal over and gently stroke its belly to "put it to sleep." When she put it down it didn't seem too sleepy, and soon wandered off.

But in that setting I wondered, as I had often before, just what American Indians made of this animal. The accounts of Indian tales and lore involving horned lizards do not say much about its third eye or its blood spraying. I'm sure that those things were noticed, and I assume explanations arose. I hope some day to hear them. What must those people, who lived in a world far richer than ours in nature-oriented spiritualism, have made of a creature that threw away its blood when all around it other animals were trying every way they knew how to keep theirs? And what did they, so often tuned to the cosmos, think of an animal that at all times kept one unblinking little eye aimed up toward the heavens?

The Elk that Wasn't

Until 1983, "Roosevelt elk" was only a name to me. I'd seen thousands of Rocky Mountain elk—often hundreds at once—and I'd read a lot about them, but I knew nothing of the three or four other surviving varieties of this glorious animal. Then in 1983 I was hired as a research consultant by the National Park Service at Mount Rainier, in Washington, as part of a large research team.

The team was investigating a variety of aspects of elk ecology in and near the park. I'm a historian by training and a naturalist by personal enthusiasm; my assignment was to examine the historical record of elk presence in the Mount Rainier area and decide if elk were indeed native to the park. So while the scientists happily hiked off to set up study transects and fly around in airplanes, I burrowed into various regional archives and libraries, emerging a few months later with an exhaustively documented list of definite maybes and a new friend. I had met the elk, and I was his.

It was easy enough to get caught up in the human romance of this animal. The ecological story, the cultural attachments, and the grand setting—the lush forests of the Pacific Northwest—contrived to bring even the dustiest manuscript to life.

Many of the native tribes of what is now Washington and

Oregon were similarly fond of elk. The Nisqually, who lived southeast of Puget Sound, had a legend that a treasure lay buried under a giant stone elk head high on the slopes of Mount Rainier. Many tribes applied elk more practically to their lives: they made clubs of the antlers, they used antler wedges to split cedar shakes, they buried their dead in elk skins, they made "parfleches" (containers of various sizes) of elk skin, and they ate elk with an enthusiasm I share today. Some waylaid elk in deep snow, other still-hunted them in midsummer in the high country. The Puyallup (pronounced *pew-AL-up*) of Washington set snares using hazel saplings on elk runways, or might simply surround an animal and beat it down. Several tribes built corrals and traps into which they could drive elk, and others drove them into bogs or lakes for easier killing. For some tribes the elk could almost approach what the bison was for Plains tribes, a walking general store of goods and services.

Elk meant much less to white settlers of Puget Sound, who by the 1840s were moving in, farming, and breeding great numbers of cattle, sheep, and horses. It didn't take long to thin out the remaining elk from the lowlands; the last one reported anywhere near Seattle was killed in 1869. The combination of developing agriculture, grazing by livestock, and overhunting had its usual effect: elk, as well as other wildlife, lost ground fast.

Oddly enough, my reading of early-nineteenth-century accounts of the region, written by a variety of explorers, adventurers, settlers, and scientists, indicated that elk numbers were already in decline when whites arrived. It remains unclear to me—and a matter of some fascination—just what was going on. No wild animal population goes along unchanged forever, but big reductions in numbers often have big reasons behind them. Something must have happened.

For example, sometimes the *influences* of Europeans—

through the introduction of trade firearms, increasing fur trade, introduction of exotic diseases from domestic stock among the elk, and so on—would reach an area years before whites actually set foot there. But no such influences seem to have had much to do with the decline of elk, so it's still a mystery to me. Perhaps climatic changes were involved.

In any case, by the late 1800s the elk were reduced to a last great stronghold, the Olympic Peninsula. Some awful travesties occurred there in the name of sport hunting, threatening to wipe out those last population remnants. Luckily, a little before the turn of the century, the outcry from legitimate sportsmen and other conservationists prevented the slaughter of these animals. The Roosevelt elk of the Pacific Northwest is not as common as it once was, but it is in no danger of disappearing.

But that's not quite true. What I should have said is that the *elk* are not disappearing. I'm not sure about the "Roosevelt" part.

The Roosevelt elk was named by Clinton Hart Merriam, one of the giants of American wildlife study, in 1897. He was then, and would be for many years, chief of the U.S. Biological Survey, and he was one of the great "splitters" of all time. Merriam was little short of incredible in his willingness to establish new species; at one point he had identified *eighty-six* species of grizzly bears in North America, primarily on the basis of minor variations in skull shape.

His most visible opponent in this tendency to split was no less than Theodore Roosevelt, an equally avid "lumper" who believed that animals, like people, varied tremendously within each species. It was a cordial disagreement, but a singularly public one. Once, when Merriam announced some new division of a game animal, Roosevelt remarked, "I have certain instincts which are jarred when an old familiar friend is suddenly cut up into eleven brand new acquaintances."

Merriam struck a virtually unanswerable blow in this verbal battle in 1897 when he decided that the northwestern form of the elk should be named for Roosevelt. In announcing the designation, he laid it on heavy: "It is fitting that the noblest deer of America should perpetuate the name of one who, in the midst of a busy public career, has found time to study our large mammals in their native haunts and has written the best accounts we have ever had of their habits and chase." This accolade was certainly no less powerful for being utterly true; Roosevelt, the amateur naturalist, could hold his own with the leading professionals of his day in his knowledge of wildlife.

But what a masterful stroke it was for Merriam. We don't know even yet if he intended it as such, but it was strategically flawless. Roosevelt was not a man to be unaffected by this sort of taxonomic immortality, and he accepted the honor gratefully. How could he do otherwise? After all, the elk was his favorite big-game animal. Other animals were named for him before his death in 1919, but none could have meant as much to him as the Roosevelt elk, even if it did cost him some points in the splitter-lumper debate.

Time told, though. Roosevelt may have lost that battle, but he won the war. Eventually, practically all of Merriam's splitting was undone. Grizzly bears were gathered back into one species, and all (or at least most) of his coyotes, moles, cougars, and the rest experienced similar reunions. After a while the Roosevelt elk, *Cervus roosevelti*, was reduced from a species to a subspecies, *Cervus elaphus roosevelti*, and now it is in danger of disappearing altogether.

One of the most exciting projects of the research team I joined at Mount Rainier was a thorough reexamination of taxonomy by a park service biologist, Christine Schonewald-Cox. Chris has just completed a study of "morphometric" characteristics, which is to say she measured a great many elk skulls from around the country to see who's who or, better put, who isn't who. She found no differences between Rocky Mountain elk and Roosevelt elk that could justify calling them separate subspecies. Another study, of genetic characteristics, agrees with her. It may take years yet, but it looks as if the Roosevelt elk is on its way out.

Too bad Roosevelt isn't around to say *Touché!* Always the eager naturalist, I suspect he would be so excited about the new scientific techniques involved that he wouldn't even mind the loss of this little bit of his immortality.

Bambi Lives

The recent release of the 1942 Walt Disney film *Bambi* in home video will introduce a whole new generation to what is still one of the most spectacular artistic successes in the history of film animation. *Bambi* has become as much a part of our culture as the fables of Aesop or the tales of Joel Chandler Harris.

Hungarian-born writer Siegmund Salzmann, writing under the pen name Felix Salten, published his novel *Bambi, A Life in the Woods* in German in 1926. It was translated into English by the American journalist Whittaker Chambers in 1928 and became a children's classic (another Salzmann book, *The Hound of Florence* (1930), became the Disney movie *The Shaggy Dog*). Salzmann's Bambi was a roe deer whose adventures occurred in a German forest. Literary historians and naturalists generally agree the novel is wonderful; it still can surprise readers with its sensitivity to ecological processes and the subtleties of wild animals' interactions with humans.

Throughout its long and profitable life, the movie *Bambi* has been criticized for its betrayal of the original story. I recognize that even the best movies rarely can do justice to the rich texture of a good novel, and that it's not fair to expect them to, but *Bambi* does deserve the criticisms it's received. As beautiful as it is, it suffers from the same homogenization that simplifies the characters in other Disney animations, such as the film version of Kipling's *The Jungle Book.* In most Disney films, it seems that in the animal world, as in the human world, there are good guys and bad guys. There are very few ambivalent guys, or troubled but essentially reasonable guys, or guys whose existence has forced upon them some vague badness that the other characters may not really understand.

The problem is not anthropomorphism itself. Anthropomorphism, which for most Disney films means the turning of nonhuman creatures into furry little people, has troubled many observers. But, as Aesop, Harris, and a host of other master storytellers have shown, giving nonhuman creatures human traits and personalities is an extraordinarily effective way to relate messages and morals, especially to young people.

What makes *Bambi* different, even today, is that it unintentionally offers lessons having nothing to do with the tidy morals of the great animal fabulists. Even though Harris's Uncle Remus tales involved wild animals, nobody would dream of trying to learn natural history or ecology from Br'er Fox. That is not how it works with *Bambi. Bambi,* probably because of its medium, teaches some really lousy ecology and has misinstructed several generations already. It would be too bad to see it happen again.

In 1988, this country experienced distress, outrage, and confusion as it watched a large part of Yellowstone National Park's forests and grasslands burn. Recent satellite mapping has shown that only about a third of the park was actually affected in any way by the flames, and less than half of that experienced the sort of intense fires that many people associate with a "destroyed" forest. But I'm not concerned here with the controversies over fire policy and fire fighting that still swirl around the park. I'm concerned with how—and why—the public met the news of the Yellowstone fires with horror while much of the scientific community, and the fire ecology specialists, did not. It has a great deal to do with *Bambi.*

The forest fire scene in *Bambi* is certainly one of the most memorable of all passages in children's film. It is terrifying. The prominent environmental historian Roderick Nash recently wrote that the movie did "more to shape American attitudes towards fire in wilderness ecosystems than all the scientific papers ever published on the subject." That deeply moving image of fire devouring everything in its path, an image we saw when we were young and our minds were virtually blank on the subject of wildlife, stuck with us.

I know it stuck with me. Even though I had read a good bit about fire ecology before I arrived in Yellowstone toward the end of the 1988 fire season, I was amazed how unlike the movie fire really is. And I was angered at how unlike the news reports the Yellowstone fires really were.

It is common knowledge among fire ecologists that fires, even the huge ones like Yellowstone's, do not annihilate all life. The fires skip around, leaving a complex patchwork that regrows as a variety of habitat types and cover types. The charred vegetation releases minerals into the soils, plants and animals (some new) colonize the burn, and nature responds vigorously to the violent ecological jolt. For important economic and social reasons, humans can't always afford to let all this happen, but it's old stuff to nature. Most primitive settings in North America were in good part shaped by periodic fire; plants and animals evolved in the face of this force, just as they evolved to deal with climatic variations and predation.

More than a decade of research by Dr. William Romme, Dr. Don Despain, and their colleagues has shown that fires on the scale of those of 1988 swept the Yellowstone region at 200- to 400-year intervals long before we were here to worry about them, with no apparent harm to the park's large wild-life populations. *Bambi* has led us astray.

The extent to which the movie was wrong is almost incredible. A team of ecological researchers led by Dr. Francis Singer recently searched the scientific literature and reported (in the journal *BioScience*) that "there is almost no published documentation of large mammals dying in either wildfires or prescribed burns in North America." Visitors and photographers who braved the smoke in Yellowstone saw elk, deer, and bison grazing or bedded down in meadows while the forest in the background blazed away.

Only on those few occasions when tremendous wind-storms whipped the fires into "runs" along a broad front could the large mammals not escape, and then most were killed by asphyxiation in the smoke rather than by burning. On those occasions, the fires were able to kill about 250 of the park's 31,000 elk, nine of its 2,700 bison, four deer (alas, Bambi), and two moose. Far more animals died the following winter because of the aftereffects of the drought and fire, and atypical climatic conditions.

But the word never really got out, did it? Dr. Conrad Smith, of the Ohio State University School of Journalism, has been conducting a comprehensive quantitative analysis of the media's handling of the fires. In a series of papers that are almost as charming as Salzmann's original novel, Dr. Smith has compiled substantial proof that "evening network news-casts misrepresented the 1988 Yellowstone fires by exaggerating their effects...." Many of us knew that, but it didn't hurt to have it shown with such scholarly rigor.

This didn't happen simply because the press loves to sensationalize things. Of course there was plenty of that, but there was also some sincere, energetic, and intelligent reporting going on. It happened, in part, because we *all* knew, even before the first lightning struck, what a fire does. We knew because we had seen Bambi and his little pals fleeing such a fire (we also knew because Smokey told us so, but that's a whole other mythos).

Such foreknowledge translated easily. A film crew would get some footage of wildlife in the park, just the kind of thing most heartily desired at the network (since you couldn't find many burning buildings or people) back in New York. Once the film arrived in New York, it was put to the best use the *Bambi*-educated producers knew.

Yellowstone's chief of research, John Varley, described one such episode: "I was with a camera crew when they filmed a running cow and calf moose. They were fleeing, all right, but they were fleeing from an incredibly loud double-rotor Veritol helicopter overhead and the activity around our vehicle. That night, I saw the footage on one of the network newscasts and, to my surprise, learned that the moose were really fleeing from 'walls of flame.' The helicopter sounds had been edited out."

Fires kill lots of trees. Fires kill some large animals, and more small animals. Fires turn green landscapes black, and some people find that exciting while others find it sickening. But fire, like animals, is easily anthropomorphized. As Bruno Bettelheim wrote in his classic study of fairy stories, *The Uses of Enchantment* (1976), "to the child, there is no clear line separating objects from living things; and whatever has life has life very much like our own." What could appear more alive than moving, climbing, swaying, roaring flames? Like the Wicked Witch of the West, or the Big Bad Wolf, the fire in *Bambi* is a villain of supernatural dimensions. But it is not a real fire.

Perhaps the biggest surprise in Salzmann's original novel is that the fire did not even happen. It was fabricated for the movie, probably to make full use of the spectacular colors involved. In the novel, the nearest thing to a supernatural villain was man, to whom the other animals always referred with a nearly deifying fear as "He" or "Him."

It is safe to say that millions of children will see the new *Bambi* video. They will meet some of American film's most beloved animated creations, and they will watch those cute, sugar-voiced little darlings flee hopelessly from the walls of flame. They will learn some ecological lessons that were discarded by fire ecologists decades ago, lessons that are not good enough in today's environmentally attuned world. It equates with learning one's racial attitudes by watching "Amos 'n' Andy" reruns, or learning how the oil industry works by watching "Dallas."

Anyone who cares about the protection of natural settings in this country knows we can't afford that kind of accidental miseducation. There is too much at stake. I imagine that many parents will sit through the *Bambi* fire scene with their children and assume that, though it's scaring the hell out of the kids, it's probably necessary for young people to know that nature *is* harsh, and life in the wild *is* violent.

But you parents should take heart; here's your chance to yield to the temptation and say something both reassuring and true, like, "Don't worry, it's just a movie; it hardly ever happens that way, and most of Bambi's little friends will get away."

Of course, I wish they'd go a little further and also say, "Bambi and his little friends would thin out pretty quickly if there wasn't a fire now and then to maintain their habitat," or even, "It's too bad they can't show you what that forest will look like a few years after the fire." But I'd be pleased just to know the kids were being reassured at all.

A Word from the Bear

Route 7 north of Manchester, Vermont, skirts the Green Mountain National Forest for several miles, winding along the upper end of the Batten Kill Valley for about five miles. Then, not long after it passes through East Dorset, it crosses an unnoticed divide and ends up in the valley of another smaller stream. There are forests in all directions, and in the summer the climate is best described as clammy.

I had been there going on five years, as Executive Director (a title of considerably more pomp than circumstance) of The American Museum of Fly Fishing, and five years was too long. I'd left work a little after six in the evening and driven north from my office, intending to go straight to my cottage in East Dorset, a village easily passed through without notice. The sullen, hazy weather was a perfect reflection of my mood.

I'd always wanted to live in New England, and the museum job was, I knew, a terrific experience that I would never forget, much less regret. But I was tired of everything eastern. I was frustrated with the impossibility of doing any of the several parts of the job as well as I would have liked; there was simply too much to do. I had started out as a curator but gradually became a fund-raiser, responsible for a series of banquets around the country and a growing amount of public relations work. It's not that I thought I was too good for that kind of work, but I didn't like it. I was, I reminded myself impatiently, a historian, after all. I may not have been an especially productive historian, but I was a lot better at being a historian than at being a salesman.

Worse, I wanted to be somewhere else. I had tried New England and was ready to go back out west. I missed naive, unfished trout that I could catch at will. I missed dry air. I missed young, sharp mountains that cast pointed shadows across sagebrush flats.

None of this was a surprise to me. I'd known all along, even the day I left Yellowstone and headed for Vermont to take the job, that I would eventually move back to the West. I don't think I realized how much the missing would ache, though. Now there were days when the car didn't seem to want to stop at my office in Manchester in the morning, when it yearned to go south an hour or so and get on Interstate 90. From there it would be simple: 2,500 miles and turn left at Livingston.

And there was nothing really stopping me. I'd already decided to leave soon, to let the museum's officers know, thank them for everything—a generous, irreplaceable everything—and drive away. Fast.

In five years I'd put out half a dozen books, all but one about the West; it was always in my mind. Some people told me I never gave Vermont a fair chance, that I was too busy writing and dreaming about the West, that the books about bears and canyons and cutthroat trout were just my way of refusing to move to Vermont.

They were probably right. Here I was in the heart of some of the best black bear country in North America, and I spent my time writing about grizzly bears. The local bears weren't cooperative—I never saw them, not at all, in my considerable odd-hour wanderings along the streams and back roads.

So there I was, driving along, fuming over all these things—an especially frustrating day at the office, the West off there somewhere beckoning to me, a breathtakingly beautiful New England valley rolling past and me unable to take it into my heart, and my usual disappointment with myself that, once again, I was too restless to settle into a truly enviable place and work there.

Before I reached East Dorset, before even its little steeples showed themselves above the trees, I decided not to go right home. I'd drive on north a few more miles and stop in to see a woman friend whom I could count on for pity and beer. This kind of funk was best savored when shared.

Oozing along the north edge of East Dorset, just past the parking lot of the Congregational church, is a little swamp, home to the occasional great blue heron and the noisiest frogs I'd ever encountered. The road passed between it and the almost immediate slopes to its west. As I hurried through the one intersection, the roof of my cottage briefly visible up the hill, the swamp became visible along the right side of the road.

I was no more than a couple of hundred yards past the general store at the junction when I caught sight of a disturbance in the shrubby trees between the swamp and the road, ahead on my right. Something was in there, and moving fast. A black form climbed from the thick tangle of saplings, up the short bank toward the road. I awoke from my daze of self-pity and discontent.

My eyes, never very trustworthy in a hurry, ventured an opinion: "A bear?"

My mind, after five years of bearlessness and in no mood for a joke, was not interested: "A dog, stupid."

My eyes got excited almost immediately. "It's a bear!"

My mind resisted, puzzled. "This is Vermont; it's a dog."

Only a second had passed now since I saw it, and my eyes were furiously relaying solid, daylight-illuminated data to my mind. The decision was made: in unison, my eyes, my mind, and my mouth agreed. "Son of a bitch, it's a bear! *It's a bear.*" There was nothing in my mind now but amazed, grateful recognition, something like I would expect to feel if I saw an old friend I had been told was dead.

It was a bear, not big but perfect, soaked and shining from a dunk in the sluggish channel that drained the swamp and had previously yielded to my sight only a few dark brook trout; a lanky black bear now rolling along, up onto the road and across, heading west with that shambling, flat-footed, yardage-eating gait that always carries them out of sight before you're ready to let them go. There he was, there he went, and there I was—just now, just as I felt the way I did about everything—and it all fit.

I heard what he said. Oh, he didn't turn his head or anything, and no one else if they had been there would have heard anything, but I heard it. Just as he crossed the road in front of my approaching car, just as he could be sure I recognized him and knew what sent him and why, I heard the message: "Go."

He was into the woods and headed up the hill by the time I stopped the car, just short of where he'd crossed the road. I hurried to the spot. His paws had left shallow dents in the gravel of the berm, and a trail of paw-slapped water curved across the pavement toward the woods. My head was still buzzing with my own shout of recognition as I bent over the nearest print. After a moment I slowly put my hand down and spread it over the print, then rose and reached with the same hand out toward where the bear had disappeared into the trees.

Acknowledgments

I must thank several fine ecologists—Don Despain, Douglas Houston, Mary Meagher, and John Varley—for their continued interest in my work over the years. Others who have been helpful in recent years, either generally or with a particular essay in this collection, are Robert Crabtree, Charles Fergus, Steve and Marilyn French, Robert Gresswell, Susan Johnston, Robert Landis, Dianne Russell, Linda Wallace, and Wade Sherbrooke.

As always, I am grateful for the help and guidance of my agent, Richard Balkin, who combines a love of the natural world with a wonderful business sense.

Many of these essays first appeared in the magazine *Country Journal,* which has had many editors. I thank Tyler Resch, John Randolph, David Sleeper, and Fred Schultz for their help at various times. Fred especially earned my gratitude, for seeing my essays into print so protectively.

Several essays also appeared in *Backpacker,* whose Tom Shealey has been a real pleasure to work with.

The publication history of the essays is as follows:

A few paragraphs from "Starker's Doves" first appeared in shorter form in an introduction I wrote to a new edition of John Crompton's *Ways of the Ant* (New York: Nick Lyons Books, 1988). "Embryonic Journeys" first appeared in *Backpacker* (October 1989), as "Stopping Nature's Clock"; "Chickadee Down" in *Country Journal* (March 1987); "Crawdad Eyes" in *Country Journal* (October 1988), as "Flight of the Crawdad"; "Antlers Aweigh" in *Country Journal* (December 1986); "The Longest Meadow" in *Country Journal* (April 1987), as "The Endless Meadow," and in *American Forests* (September–October 1987); "Keep Moving" in *Country Journal* (December 1988), as "Feeding Frenzy"; "Songs of the Border" in *Country Journal* (August 1988); "Getting the Drift" in *Country Journal* (August 1987); "The Odd Couple" in *Country Journal* (October 1987); "All Together Now" in *Country Journal* (June 1987); "Biological Storms" in *Country Journal* (June 1988), and in *Backpacker* (September 1988), as "Mother's Mess"; "Apologies to the Salmon" in *Backpacker* (January 1989); "Taming the Giant Weasels" in *Country Journal* (April 1988), as "Ferocious Beasts," and in *Backpacker* (July 1988), as "The Beast Within"; "A Lot of Browns and Yellows" in *Country Journal* (November 1987), as "Wildlife on Stamps"; "So Long, Sucker" in *Backpacker* (December 1989), as "The Sucker"; "Microdragons" in *Country Journal* (December 1987); "The Elk that Wasn't" in *Country Journal* (February 1988), as "Lumping the Elk"; "Bambi Lives" in *Backpacker* (November 1990).

THE MOUNTAINEERS, founded in 1906, is a non-profit outdoor activity and conservation club, whose mission is "to explore, study, preserve and enjoy the natural beauty of the outdoors ..." Based in Seattle, Washington, the club is now the third largest such organization in the United States, with 12,000 members and four branches throughout Washington State.

The Mountaineers sponsors both classes and year-round outdoor activities in the Pacific Northwest, which include hiking, mountain climbing, ski-touring, snowshoeing, bicycling, camping, kayaking and canoeing, nature study, sailing, and adventure travel. The club's conservation division supports environmental causes through educational activities, sponsoring legislation, and presenting informational programs. All club activities are led by skilled, experienced volunteers, who are dedicated to promoting safe and responsible enjoyment and preservation of the outdoors.

The Mountaineers Books, an active, non-profit publishing program of the club, produces guidebooks, instructional texts, historical works, natural history guides, and works on environmental conservation. All books produced by the Mountaineers are aimed at fulfilling the club's mission.

If you would like to participate in these organized outdoor activities or the club's programs, consider a membership in The Mountaineers. For information and an application, write or call The Mountaineers, Club Headquarters, 300 Third Avenue West, Seattle, Washington 98119; (206) 284-6310.